EXTINCT ANIMALS

EXTINCT ANIMALS

An Encyclopedia of Species That Have Disappeared during Human History

Ross Piper

Illustrations by Renata Cunha and Phil Miller

GREENWOOD PRESS

Westport, Connecticut • London

Library of Congress Cataloging-in-Publication Data

Piper, Ross.
    Extinct animals : an encyclopedia of species that have disappeared
during human history / Ross Piper ; illustrations by Renata Cunha and
Phil Miller.
        p. cm.
    Includes bibliographical references and index.
    ISBN 978–0–313–34987–4 (alk. paper)
    1. Extinct animals—Encyclopedias.   I. Title.
    QL83.P57   2009
    591.6803—dc22   2008050409

British Library Cataloguing in Publication Data is available.

Library of Congress Catalog Card Number: 2008050409
ISBN: 978–0–313–34987–4

First published in 2009

Greenwood Press, 88 Post Road West, Westport, CT 06881
An imprint of Greenwood Publishing Group, Inc.
www.greenwood.com

Printed in the United States of America

The paper used in this book complies with the
Permanent Paper Standard issued by the National
Information Standards Organization (Z39.48–1984).

10 9 8 7 6 5 4 3 2 1

We live in a zoologically impoverished world, from which all the hugest, and fiercest, and strangest forms have recently disappeared.

—Alfred Russel Wallace (1876)

To my Mum, Gloria

CONTENTS

PREFACE

Countless books have been written about the dinosaurs, the reptiles that ruled the earth for about 160 million years, yet remarkably few books have been written about the many strange, fierce, and enormous beasts that have disappeared in the time humans and our recent ancestors have been around. The earth is certainly a poorer place for their passing, but it's fascinating to think that our forebears knew these animals—even worshipped them and hunted them. *Extinct Animals* is an exploration of these creatures, from the giant, flesh-eating birds and saber-toothed marsupials of South America to the golden toad of Costa Rica, which became extinct as recently as 1989.

A book on extinct animals would not be complete without a little about the process of extinction itself, and so in the introduction, you find out about how the earth has been rocked by numerous mass extinction events. The last of these, the seventh extinction, is happening right now as a result of the unchecked growth of the human population and the habitat destruction that follows in the wake of what we call progress. Following the introduction are 65 vignettes, each of which present a different extinct animal. You will not find an exhaustive account of all the animals that have disappeared from our planet in the last couple of million years because such a book would be immense, and all that we know of many extinct animals is based on fragmentary fossils. The focus of this book is those extinct beasts for which there are historical accounts of the living animal, a detailed fossil record, or scant remnants that indicate a truly incredible creature.

The audience for *Extinct Animals* is anyone with an interest in zoology, earth's remarkable recent past, or the far-reaching consequences of an expanding human population. The main purpose of *Extinct Animals* is to present what we know about the lives of animals that have disappeared forever in a way that just about anyone can read and understand. Textbooks are full of fascinating information, but all too often, they are inaccessible to general audiences. This book provides a bridge to those resources for anyone who has even the slightest inter-est in the world around him and what it was once like.

Along with the individual vignettes are a number of entries that describe some of the discoveries and concepts that are crucial to understanding how life on earth has changed in the last couple of million years. These include the amazing bone deposits of Rancho La Brea in Los Angeles, the ice ages, and the human age of discovery, which has seen humans exploring every corner of the globe, often to the detriment of native fauna.

Wherever possible, I have tried not to use jargon. There is a whole dictionary of specialized zoological and paleontological terms, which can sometimes be confusing or difficult to say. I have tried to write in more general terms without using this specialized language. However, there is a glossary at the end of the book to explain any jargon that was unavoidable. For those readers keen to trawl the Web for extra information, the best way is to type the Latin name, or perhaps the common name, into a Web search engine. The amount of information on the Web today is such that there will be numerous pages on most of the animals in this book, but only those sites ending in .gov or .edu are likely to carry information that has been thoroughly researched and edited.

In this book, at the end of many entries, there is a list of resources for further reading. These lists, as well as the selected bibliography at the end of the book, include textbooks and journal articles that can be found in any decent library. In addition to the Web and books, you can find more about the animals featured in this book by visiting natural history museums. A list of some of the museums where you can see skeletons and reconstructions of many extinct animals can be found at the back of this book.

ACKNOWLEDGMENTS

I would like to thank the following people who have helped me with this book by reviewing content and providing me with photographs: Robert McNeill Alexander, Herculano Alvarenga, Christine Argot, Jennifer Rae Atkins, Susana Bargo, John Clay Bruner, Per Christiansen, Russell L. Ciochon, Darin Croft, Matt Cupper, Chris Dickman, Judith Field, Claude Guérin, Michael D. Gottfried, Tim Halliday, Fritz Hertel, Don Hitchcock, Christine Janis, Paul Johnsgard, Paul Kitching, Rob Kruszynski, Tatiana Kuznetsova, David Laist, Roger Lederer, Adrian Lister, Jeffrey Lockwood, Marco Masseti, Raoul Mutter, Pancho Prevosti, Julian Pender Hume, Víctor Hugo Reynoso Rosales, Dave Roberts, Hans Rothauscher, John D. Scanlon, Elwyn L. Simons, Nikos Solounia, John D. Speth, Mary Stiner, Tony Stuart, Ian Tattersall, Eduardo Pedro Tonni, Cis van Vuure, Ashley Ward, Rod Wells, Richard S. Williams, Paul Willis, and Michael Wilson.

The following institutions have also kindly provided me with photographs: the Natural History Museum at Tring, the Royal Saskatchewan Museum, the Texas Memorial Museum, and the Australian Museum.

Finally, I would like to say a big thanks to Renata Cunha and Phil Miller for the excellent illustrations you will see throughout this book.

INTRODUCTION

Extinction is a fact of nature. All of the species of animal that live on earth will, at some point, become extinct. Some, through the process of evolution, may give rise to descendents—new forms to exploit different niches—while others may disappear, leaving no line of descent.

Ever since animals made their first appearance in the story of life on earth, billions of species have disappeared. Some of these have fallen to some huge, cataclysmic events, of which there have been many in the last few hundred million years, while others have been outcompeted by other organisms or were unable to react to small changes in their environment. In 1982, scientists proposed that in the last 500 million years—a window of time in which animals have evolved to exploit the vast majority of habitats on earth—there have been around six mass extinction events. It's fascinating to think how life on earth has been pushed to the edge on a number of occasions, only to spring back with renewed vigor when conditions have become more favorable.

These mass extinctions happened such a long time ago that the evidence for what caused them is not immediately obvious, and for some of them, the evidence may have been worn away completely. Scientists have attributed these extinctions to meteorite impacts, massive volcanic eruptions, and movement of the solar system through a galactic gas cloud, to name but a few explanations. Regardless of the cause, some of these events saw the disappearance of huge numbers of species. The largest of these mass extinctions, which occurred 250 million years ago, resulted in the disappearance of 96 percent of all marine life and around 70 percent of all terrestrial life. During this time, animal life must have been pushed to the very edge, reduced to a shadow of its former glory—perhaps a few species clinging on to life in what had become a very harsh world indeed. We may only be able to guess at the causes of these extinctions, but the fossil record gives us a glimpse of these times. To those who can decipher it, the fossil record from around these periods shows an unprecedented die-off of species, with many disappearing completely. The fossil record is a story in stone, shell, and bone of life on earth. It enables us to picture the lives of long-dead creatures and shows how cataclysmic events have ravaged life on earth on numerous occasions.

THE MAJOR EXTINCTIONS AND THEIR CAUSES

Cambrian-Ordovician

Geologists use a series of extinction events that occurred around 490 million years ago to define the end of the Cambrian period and the beginning of the Ordovician. These events led to the demise of many types of marine animal. The brachiopods (marine mollusks resembling bivalves) were very numerous before this event, but whatever occurred all that time ago had a drastic effect on their numbers. The trilobites, ancient forerunners of today's numerous creepy crawlies, could also be found in profusion before this event, but the Cambrian-Ordovician mass extinctions heralded a slow decline of these organisms that lasted for millions of years.

What caused this series of extinction events almost 500 million years ago? No one can be sure, but many scientists suggest it was a lengthy series of glaciations. By far the most important source of energy for life on earth is the sun. Its heat, reaching out over millions of miles of space, ensures that the earth has a balmy climate—well, some of the time. The problem is that our planet does not travel around its star in a perfect orbit. There are rhythmic variations, not only in how the earth goes around the sun, but also in the way the earth spins on its axis (see the "Extinction Insight" in chapter 5 for more information). All of these anomalies have a huge effect on the earth's climate. For example, small wobbles in the earth's spin can reduce the amount of solar radiation that strikes the Northern Hemisphere, and temperatures can drop—not by a massive amount, but enough to result in the formation of huge glaciers that can lock up much of the planet's water. These ice ages, as they are known, have a huge effect on the earth's inhabitants, which is not surprising as life generally fairs better in a greenhouse than in a refrigerator. It has been suggested that a sequence of ice ages was responsible for the Cambrian-Ordovician extinction events. Life at this time was at its most diverse in the numerous shallow seas that surrounded the earth's landmasses. If the earth did indeed enter a long phase of ice ages, the water from these shallow lagoons and seas disappeared as the world's moisture was locked up in the growing glaciers.

Another possibility is that the action of bacteria living in the mud on the sea floor led to the depletion of oxygen in the ocean, which in itself is due to climate change. All animal life at this time was marine—there were no land-dwelling creatures—and all animals require oxygen. Deprived of oxygen, animals would have gone into a steep decline.

Ordovician-Silurian

In scarcely no geological time at all after the Cambrian-Ordovician mass extinction, the fossil record tells us that there was another big die-off of species around 450 million years ago. It is likely that this Ordovician-Silurian mass extinction was also a series of events which occurred quite close together—in geological time, at least. This mass extinction is widely considered to be the second largest the world has ever seen, and it resulted in the loss of around 50 percent of the animal types that were around at that time.

Again, we can only make educated guesses at the culprit, but climate change is a definite possibility, such are the vagaries of earth's motion through space. A series of ice ages and warmer periods led to the cyclical rise and fall of sea levels. Before this series of changes, the shallow seas would again have been the focus of animal activity, but global cooling deprived these creatures of the habitat they required. In the intervening warm periods, the creatures that evolved to live in the new habitats provided by the cool conditions were doomed. So this cycle continued for hundreds of thousands, if not millions, of years, with animal diversity decreasing all the while.

Devonian-Carboniferous

The late Devonian extinction happened around 360 million years ago, and instead of one event, it seems the decline in animal species, which marks the beginning of the Carboniferous period, was also a series of events that lasted for around 20 million years. Again, we can never be sure of the underlying causes that resulted in the loss of around 70 percent of all species, but numerous theories have been suggested, including a large asteroid impact and the evolution of the plants from small, surface-hugging forms, no larger than 30 cm, to giants 30 m tall. These new plants had well-developed roots that penetrated bedrock and led to the eventual formation of thick layers of soil. Rainwater running through this soil carried huge quantities of minerals to the sea, completely changing its chemistry and creating algal blooms, which sucked the oxygen out of the water. Starved of oxygen, marine animals perished. This is just a theory, but it is an event that could have conceivably been played out over millions of years. The profusion of land plants may have also caused extended periods of glaciation by removing carbon dioxide, an important greenhouse gas, from the atmosphere.

Permian-Triassic

The mother of the major extinctions is the one that occurred at the end of the Permian period (about 250 million years ago), an event which defines the beginning of the Triassic. The Permian-Triassic extinction killed off around 96 percent of all marine species and about 70 percent of land-dwelling species. Many theories for the cause of this event have been suggested, and some are more credible than others. The usual suspect of an asteroid strike (or even multiple strikes) has been proposed, but in the absence of definite crater(s), we cannot be sure if this was the case. At around the right time to coincide with the Permian-Triassic event, there appears to have been a massive increase in volcanic activity. The Siberian Traps are the lasting reminders of this colossal outpouring of basalt from the earth's mantle. In this scenario, a plume of hot magma from the deep mantle rose up and ruptured the crust, appearing as a series of eruptions over a huge area. Eruptions of this type do not end after a few days, and it appears that the basalt of the Siberian Traps was spewed out over a million years. Imagine all the dust and gases ejected into the atmosphere by an outpouring of 3 million km^3 of lava (for comparison, the largest eruption in very recent history occurred in Iceland, and it produced 12 km^3). Mount Pinatubo ejected only a tiny fraction of the gases and dust produced by the Siberian Traps eruptions, yet this was enough to lower global temperatures by 0.5 degrees Celsius, which is not an inconsiderable drop for living things.

Imagine how the earth's climate was cooled by the Siberian eruptions. The effect must have been like a nuclear winter, and photosynthesizing organisms, the basis for all food chains on land and most in the oceans, died en masse. Huge amounts of noxious gases pouring into the atmosphere acidified the moisture in the air, and thousands of years of global acid rain made the oceans more acidic, dissolving corals and countless other organisms that secrete a shell of calcium carbonate. As the eruption occurred over an area the size of Europe, molten rock heated seawater, and immense storms may have formed. These hypercanes, with winds in excess of 800 km per hour, sucked dust, debris, and gases into the high atmosphere, eroding the ozone layer until the earth was stripped of its protection from ultraviolet radiation.

We know that this huge, prolonged volcanic eruption occurred in Siberia about 250 million years ago, but there is a possibility that it may have been triggered by a huge asteroid impact. An errant cosmic body, bigger than the largest mountain, slamming into the earth at 15 to 20 km per second generates an unimaginably huge amount of energy. Is this enough to disturb the currents of molten rock that flow through the earth's mantle, causing the creation of a gigantic plume of molten rock that bursts from the surface and wreaks millions of years of havoc? Possibly, but until we find the remnants of a crater of the right age and size, the trigger of the Siberian eruption will remain a mystery.

Another very interesting proposed cause of the mass extinction at the end of the Permian is the release of huge quantities of natural gas from below the seabed. Beneath the seabed, this gas (mostly methane) is locked away within the crystal structure of frozen water, and a huge impact or an increase in ocean temperatures due to a colossal eruption may have been enough to melt these extensive reserves, releasing huge quantities of methane into the atmosphere. Methane is one of the most potent greenhouse gases of all, and billions of tonnes of it released all at once could have triggered a runaway greenhouse effect that turned the earth into a sweltering sphere for thousands of years. Any one of these events (flood eruption, asteroid impact, or an enormous release of methane) would be very bad news for all life, but if all three were perhaps linked, it must have been as close to the end as life has ever come.

Triassic-Jurassic

The next mass extinction after the Permian event is the one that divides the Triassic period from the Jurassic: the Triassic-Jurassic extinction event. This was minor compared to the event that went before it, but it is significant enough to have been preserved in the fossil record, with the disappearance of many marine forms as well as a range of land animals. Some scientists have challenged whether this was actually a real event or just a reduction in the appearance of new species. An asteroid impact has been proposed as a possible cause, but no crater of the right age or size has been found. This is definitely not the case for the next major extinction, and perhaps the most famous of them all, for it is when the dinosaurs disappeared from the earth.

Cretaceous-Tertiary

Dinosaurs have fascinated us since the first species was described in 1824, yet almost all of them disappeared rather abruptly around 65 million years ago along with countless other species. I say almost all because birds are the direct descendents of these animals. Every time you look at your bird feeder or see a flock of geese heading south for the winter, you are seeing the lasting reminders of these reptiles. What happened to the rest of the dinosaurs? The event that ended their dominance is known as the Cretaceous-Tertiary extinction event, or the K-T event for short (Cretaceous is traditionally abbreviated as *K*, derived from the German word for chalk, *kreidezeit*), and it is the only mass extinction for which there is definite evidence of an asteroid impact. In numerous sites around the world, geologists saw that rock strata laid down in the Cretaceous were topped off with a thin layer of grayish material. This layer turned out to be ash, and further analysis showed that it contained a high concentration of the very rare metal, iridium. Iridium may be rare on earth, but it is much more abundant in certain types of asteroid. For years, skeptics argued that the iridium could have originated deep in the earth's mantle and been ejected by intense volcanic activity. Also, they argued, how could there have been an asteroid impact with no crater? Then, in 1990, geologists formally identified the crater from observations made many years before. The site is known as Chicxulub, and it is on the very edge of the Yucatan Peninsula in Mexico. The crater is half on the land and half under the sea, but after 65 million years, the portion on land has been eroded and the submarine half is buried under hundreds of meters of sediment. With that said, it is possible to get an idea of its size, and it is truly immense, with a diameter as large as 300 km. The space rock that formed this crater was at least 10 km across and was traveling at around 15 to 20 km per second. Such an enormous thing hitting the earth at such a high speed generated a huge amount of energy—at least 2 million times more energy than the largest nuclear bomb ever detonated. Huge waves ravaged the earth's low-lying areas, and the huge quantities of dust and gas ejected into the atmosphere plunged life into darkness for months, if not years. With insufficient sunlight, plants and other photosynthesizing organisms everywhere died, and the animals followed. Some geologists have suggested that the earth was hit by several asteroids around 65 million years ago, but the other craters are yet to be found. As most of the planet is covered by water, lots of impact craters may be buried beneath the waves and hundreds of meters of sediment.

At around the same time as the Chicxulub crater was formed, the earth was struck by a second terrible event, a second huge volcanic flood eruption that produced the Deccan Traps in India. Again, it's feasible that the impact triggered an outpouring of basalt on the other side of the world, and the effects of both together spelled disaster for all life.

THE ANCIENT MASS EXTINCTIONS AND EVOLUTION

For the organisms that experience them, cataclysmic events bring death and devastation, but mass extinctions have their positive side, too. Indeed, if it wasn't for mass extinction, we would not be here. Mass extinctions wipe the biological slate clean and leave the door open to organisms that have been kept in the shade. If we travel back in time, the Permian-Triassic mass extinction created an opportunity for the dinosaurs to rise to dominance, following the demise of the large synapsids, i.e., *Edaphosaurus*, *Dimetrodon*, and so on. This is known as the Triassic takeover, and as the dinosaurs diversified and grew larger, the surviving synapsids were forced into the shadows as nocturnal, insectivorous animals, and they gradually evolved the characteristics that we know as mammalian. For 160 million years, these animals and their true mammal descendents lived in the shadows of the dinosaurs, scurrying around the feet of the reptilian giants. Then, 65 million years ago, the K-T event ended the dominance of these reptiles and the door was wide open. For a short while after the disappearance of the dinosaurs, the mammals, birds, and crocodiles all vied with each other to take the place of the extinct reptiles. Eventually, the mammals were successful, and they evolved remarkably quickly to fill the niches in the post dinosaur world.

THE SEVENTH EXTINCTION

As strange as it may sound, we could actually be living in the middle of a mass extinction right now. In recent times, there have been no colossal outpourings of lava, nor have there been any huge asteroid impacts, so what's the cause of this, the seventh, mass extinction? We are. Humans almost certainly contributed to the demise of some of the Pleistocene animals, some of which appear in this book. More recently, around 780 species have become extinct since 1500, but as the vast majority of species disappear without us knowing anything about it, the real number is far higher. Scientists estimate that during the last century, somewhere in the region of 20,000 to 2 million species became extinct, and in the next 100 years, humanity's wholesale destruction of habitats around the globe could result in the extinction of 50 percent of all species.

The problem is that the human population is growing out of control. In around 8000 B.C., the human race numbered around 5 million individuals. In 1750, there were around 750 million people, but today, there are around 6.6 billion of us. At the moment, the human population grows by 76 million people every year. Imagine trying to find living space, food, and water for all those people. Also, better health care means that the population growth is accelerating. As the human population grows, more and more pressure is placed on the natural world. We destroy natural ecosystems to make space for our crops and buildings, and yet more pristine habitats are ruined by the poisonous products of our agriculture and industry. The tropical rainforests are the most biologically diverse habitat on the planet. They cover only 2 percent of the earth's surface, yet they are home to 50 percent of all living species. They are so rich in life that a single rainforest tree may be home to several species of plant, animal, and microbe found nowhere else on earth, but with every passing year, they are being burned and chopped down. Every second that passes sees the loss of one and a half acres of tropical rainforest, and if the present rate of destruction continues, the tropical rainforests will be consumed in 40 years, with tragic consequences for every living thing on the planet.

Like the tropical rainforests, the world's oceans teem with life, but the condition of the marine ecosystem is now nothing less than a global emergency. Huge fleets of fishing vessels haul millions of tonnes of fish, crustaceans, and mollusks out of the water every year, and many stocks of commercial species have collapsed completely because of this relentless and senseless hunting. Millions of liters of toxic effluents, dangerous wastes, and agricultural run-off make their way into the ocean every year, and in some places, these have already killed off much of the marine life.

We have no idea how many species of organism live in the world's most biodiverse places, and with every passing year, species become extinct before we even knew they existed. Until we understand that we are one species among many and that our continued survival depends on living in harmony with the natural world, the future looks very bleak for the human race and the other species with which we share this planet.

FEWER THAN 100 YEARS AGO

GOLDEN TOAD

Golden Toad—The golden toad was restricted to the cloud forest above the city of Monteverde in Costa Rica. It was last seen in 1989. (Renata Cunha)

Scientific name: *Bufo periglenes*
Scientific classification:
Phylum: Chordata
Class: Amphibia
Order: Anura
Family: Bufonidae
When did it become extinct? No golden toads have been seen since May 1989.
Where did it live? The golden toad was only known from an area of cloud forest above the city of Monteverde in Costa Rica.

The disappearance of the golden toad was both mysterious and rapid. Only 25 years separate the species' discovery by scientists in 1964 and the last sighting in 1989. Since its disappearance, this 5-cm-long toad has become an icon for the decline of amphibians the world over.

Unlike the majority of toad species, the male golden toad was brightly colored and shiny to the extent that it looked artificial. The species was also unusual as the male and female were very different in appearance. The male, with his magnificent golden orange skin, was in stark contrast to the larger female, who was black with scarlet blotches edged in yellow.

This toad was only known from a small area (around 10 km²) of high-altitude cloud forest in Costa Rica that today is part of the Monteverde Cloud Forest Reserve. These forests (also known as elfin forests) are characterized by cloud, epiphytic plants galore, and small trees, which all in all give them a very primeval feel. In this small area of perpetually moist forest, the golden toad could apparently be encountered commonly and in large numbers, but only during the breeding season. The breeding season extended from April to June, when the rainy season is usually at its most intense. These rains would fill the hollows around the bases of trees and other natural depressions with water—ideal toad breeding pools. The toads would collect around these pools in great numbers with the sole intention of breeding. Mating in any toad species is far from genteel, and golden toad breeding was a free for all. The males outnumbered the females by eight to one, and any female in the vicinity of a breeding pool soon found herself beneath a writhing mound of potential suitors. The males would get so excited and desperate that they would try to mate with anything that moved, including other males. Occasionally, between 4 and 10 feverish males would grab hold of each other to form a toad ball the significance of which is unknown—perhaps a female was in the middle of the ball but managed to give her suitors the slip. Once a male had struggled with his competitors and beaten them to get a good hold of a female in the breeding grasp known as amplexus, he could fertilize her eggs—or at least, this was his intention. Often, other males would come along and try to separate the mating couple so that they could get a chance at fertilizing the female's eggs.

What with all this wrestling and bad sportsmanship, it's quite surprising that the golden toad managed to breed at all, but breed they did, and the female would eventually lay 200–400 3-mm eggs in a long string in the breeding pool. Compared with many species of toad, the golden toad laid relatively few big, yolk-packed eggs, rather than lots of small ones, and it is thought this breeding strategy evolved because of the small size of the pools on which the toad depended. These pools didn't last very long, and so after the tadpole hatched, the race was on to change into a toadlet as quickly as possible. The abundant yolk in the eggs was the fuel for this rapid development.

After hatching, the tadpoles would spend around five weeks in the ephemeral pools before they lost their tadpole features and sprouted limbs, enabling them to begin their life on land. What the toads did outside of the breeding season is unknown. We don't know what food they ate and how they went about catching it. The adults of the majority of other toad species are pretty unfussy when it comes to food, and they go for just about any creature that will fit inside their capacious mouth. There is no reason to believe the golden toad was any different, but its small size restricted it to small animals like insects and other invertebrates.

Like much of the golden toad's biology, we also have a poor understanding of why it disappeared. We know that when it was first discovered by Western scientists in 1964, it was found in large numbers, but in a very small area. In 1987, 1,500 adults were seen, but then in both 1988 and 1989, only one adult was seen. What happened to cause such a massive population crash? We don't know for sure, but there are three main theories. It has been suggested that as the toad had such special breeding requirements—short-lived pools and a narrow window of opportunity—one erratic year of weather conditions would have completely scuppered their chance of a successful breeding season. Species like the golden toad have very specific habitat requirements, occupying very small ranges. This predisposes them to extinction as one little change in their environment can leave them with nowhere to go. Other scientists have suggested increasing amounts of ultraviolet (UV) radiation penetrating the atmosphere could have harmed the toads, but as they lived in dense forest shrouded in cloud during the breeding season, this is unlikely to be the cause of their demise. The last theory concerns the spread of chytrid fungi, which appear to make short work of amphibian populations wherever they become established. Drier conditions could have forced the toads into fewer and fewer ponds, increasing the transmission of this disease. With this said, it is possible that the golden toad still clings to existence in some remote corner of Central America.

+ The cloud forests of Monteverde have lost 40 percent of their frog and toad species, and it is not only here that amphibians are in trouble. In the past three decades, scientists all over the world have reported massive declines in amphibian populations, with some 120 species thought to have become extinct since the 1980s. The declines and the extinctions are global, but the United States, Central America, South America, eastern Australia, and Fiji have been worst hit.

+ Chytrids, a group of pathogenic fungi, are often blamed for this decline. This disease was first noted on a captive frog in Germany, but its global spread has been linked to the trade in the African clawed frog, an animal that is used in laboratories the world over for a plethora of experiments. American bullfrogs have also spread around the world thanks to the pet trade, and these, too, carry the chytrid fungi, although they are not affected by the disease.

+ Although the chytrids do cause disease and death in amphibians, it is unlikely they are wholly responsible for the global decline of these animals. There are probably numerous factors at play, including habitat destruction, climate change, and increasing levels of UV radiation. Only intensive research will allow us to solve the puzzle and halt the decline of these interesting animals.

Further Reading: Savage, J.M. "An Extraordinary New Toad from Costa Rica." *Revista de Biología Tropical* 14 (1966): 153–67; Jacobson, S.K., and J.J. Vandenberg. "Reproductive Ecology of the Endangered Golden Toad (*Bufo periglenes*)." *Journal of Herpetology* 25 (1991): 321–27; Phillips, K. *Tracking the Vanishing Frogs.* New York: Penguin, 1994.

GASTRIC-BROODING FROG

Scientific name: *Rheobatrachus silus*
Scientific classification:
    Phylum: Chordata
    Class: Amphibia
    Order: Anura
    Family: Myobatrachidae
When did it become extinct? This frog was last seen alive in 1981.
Where did it live? The gastric-brooding frog was known only from the Canondale and
    Blackall mountain ranges in southeast Queensland, Australia.

Another victim of the amphibian disaster was a fascinating little frog from Australia that was only discovered in 1973, yet by 1981, it had vanished without a trace.

The gastric-brooding frog was a small species; females were around 5 cm long, while males were smaller, at approximately 4 cm. It lived in forest streams and rocky pools, and for much of the time, it would hide beneath rocks on the bed of these water bodies, but when

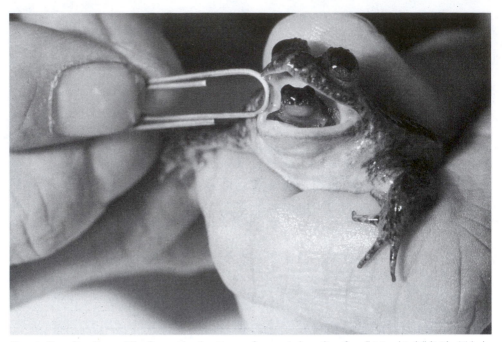

Gastric-Brooding Frog—The first and only picture of a gastric-brooding frog "giving birth." (Mike Tyler)

it left these rocky refuges and moved out into the fast-flowing water, it showed itself to be a very accomplished swimmer. Its powerful hind-limbs terminated in feet that were almost completely webbed, and these were used with good effect to propel the frog through the water. The big, protruding eyes of this frog were positioned well on top of its head, and this allowed it to survey what was going on in the air and on land, while its body was out of sight beneath the water. Although it was very well adapted to an aquatic existence, the gastric-brooding frog would often leave the water to hunt or to seek out a new stretch of stream.

Its favored prey were small invertebrates, such as insects, but unlike many types of frog, the gastric brooder did not have a long, sticky tongue to secure its prey; instead, it waited until its food was within range and simply lunged at it with an open mouth. With its prey partially trapped, the frog would shove the rest of the victim's body into its mouth using its forelimbs. Even though this frog was a capable predator, it was very small, and it was a tasty morsel for a range of predators. Herons and eels were partial to this amphibian, but it did have a useful defense if it was grabbed by one of these animals: mucus. All amphibians have skin glands that produce mucus to keep their skin moist as well as for protection. The gastric brooder could produce lots of very slippery mucus, which made it very hard for a predator to get a good grip.

In most respects, the gastric-brooding frog was like most other frogs, but what set it apart was the way it reproduced. Mating was never observed in this species, but it is known that the female laid between 26 and 40 eggs and that these were then fertilized by the male. Again, this is the normal amphibian approach when it comes to breeding as fertilization in all these animals is external. It is not completely clear what happened next as it was never actually seen, but at some point after the eggs were fertilized, either when they were still eggs or when they had hatched into tiny tadpoles, the female swallowed as many of her off-spring as she could. To the uninitiated, this may have looked like maternal cannibalism, but in fact, this was part of this frog's unique reproductive strategy. The eggs or small tadpoles slipped down their mother's throat and ended up in her stomach, and this is where they grew. In all animals, the stomach is the organ that plays a major role in digestion. Cells in the lining of the stomach produce very strong acid that breaks down food into its compo-nent fats, proteins, and carbohydrates so that enzymes can begin their digestive work. This harsh, acidic environment is hardly ideal for developing offspring, but over millions of years, these frogs evolved a couple of tricks that turned the stomach into a snug little capsule for their developing brood. It seems that the eggs and the tadpoles of this frog secreted a type of chemical known as a prostaglandin. This chemical blocked the cells of the stomach lin-ing from secreting acid, and the walls of the stomach thinned. The young frogs turned the stomach into a cozy crèche. After six to seven weeks of developing in their mother's alimen-tary canal, 6 to 25 tiny but fully developed froglets clambered out of their mother's mouth to begin their own life in the big wide world. Throughout this whole brooding period, with her stomach effectively shut down, the female frog was unable to feed, so after the departure of her young, her first consideration was probably finding some food.

In fewer than 10 years after its discovery, the gastric-brooding frog disappeared. Exten-sive searches of the mountain streams in the early 1980s failed to turn up a single specimen. When the species was first discovered in 1973, it was considered to be quite common, but by 1981, not a single specimen was to be found—it was as though it had been spirited away. Like the golden toad of Costa Rica, exactly what happened to the gastric-brooding frog is unknown, but there have been several explanations, some of which are more plausible than others. Pollution of the mountain streams by logging companies and gold panners has been cited as a reason for the disappearance of this species, but tests on the stream water failed to show any significant pollution. Habitat destruction has also been mentioned, but the areas where this frog was found have been pretty well protected. With pollution and habitat destruction largely ruled out, we arrive at the specter of disease. The chytrid fungus has caused the deaths of amphibians all over the world. The fungus latches on to the body of an amphibian and takes root in its skin. The fungus forms cysts within the deeper layers of the skin and breaks down keratin, a protein in the cuticle of many vertebrates, including adult frogs and toads. The skin of an amphibian infected with this fungus begins to break down, and in severe cases, the disease can eat right into the deeper tissues. In these cases, digits, and even limbs, can be eaten away. This in itself is not fatal, but the ability of the skin to transport gases and prevent the entry of other harmful micro-organisms is probably im-paired, and the victim dies a slow and probably very painful death.

+ The species discussed here is actually the southern gastric-brooding frog. In 1984, a very similar species, the northern gastric-brooding frog (*Rheobatrachus vitellinus*), was discov-ered living in the Clarke Mountains near Mackay in central coastal Queensland. A year later, this species also suffered a total population crash, and it has not been seen since.
+ The gastric-brooding frog was very vulnerable to extinction as its range was so small. It existed in one small corner of Australia and nowhere else on earth.

+ The chytrid fungus is not native to Australia, but it has somehow been transported there either by the pet or laboratory animal trade. The gastric-brooding frog probably had little or no immunity to the chytrid fungi. In a situation like this, a disease-causing organism can spread very rapidly indeed.
+ In Darwin's frog, the tadpoles develop in the vocal sacs of their father, a strategy that doesn't involve periodic starvation like gastric brooding.

Further Reading: Corben, C.J., M.J. Ingram, and M.J. Tyler. "Gastric Brooding: Unique Form of Parental Care in an Australian Frog." *Science* 186 (1974): 946–47; Tyler, M.J., D.J. Shearman, R. Franco, P. O'Brien, R. F. Seamark, and R. Kelly. "Inhibition of Gastric Acid Secretion in the Gastric Brooding Frog *Rheobatrachus silus*." *Science* 220 (1983): 607–10.

ESKIMO CURLEW

Scientific name: *Numenius borealis*
Scientific classification:
   Phylum: Chordata
   Class: Aves
   Order: Charadriiformes
   Family: Scolopacidae
When did it become extinct? The Eskimo curlew is thought to have become extinct around 1970.
Where did it live? In the northern summer, the Eskimo curlew spent its time in the Canadian subarctic. Its wintering grounds were the Argentinean Pampas, south of Buenos Aires.

Eskimo Curlew—In addition to existing in huge numbers, the Eskimo curlew annually tackled one of the most arduous migrations in the natural world. (Renata Cunha)

The story of the Eskimo curlew is a sad tale of greed and senseless waste and a perfect example of how destructive our species can be. The Eskimo curlew was a small wading bird, no more than 30 cm long, with an elegant, 5-cm-long beak. Like the other curlew species, the Eskimo curlew had a distinctive, beautiful call, and the Inuit name for this bird, *pi-pi-pi-uk*, is an imitation of the sound they made on the wing and on the ground.

The Eskimo curlew may have been a small bird, but it was one of the most accomplished globetrotters that has ever graced the skies. Like many other species of wading bird, this curlew spent its time between northern breeding grounds and southern wintering grounds. Traveling between the two was no mean feat, and the small birds had to embark on one of the most complex and dangerous migrations in the animal kingdom. As the short, northern summer ended and the curlew's young had been reared, the birds took to wing for the beginning of an arduous and dangerous journey. Its migration took it in an immense clockwise circle, starting from the subarctic Canadian tundra, through the Western Hemisphere and east through Labrador, down through the Atlantic and across the southern Caribbean. The birds continued this epic journey until they reached their wintering grounds on the Argentinean Pampas. Some of the migrating birds went even further, eventually reaching Chile. The birds would spend a few months in South America until the spring returned to the north and the pull of hundreds of thousands of years of habitual behavior forced them into the air, en masse, for the return leg. The return to the breeding grounds took them through Texas, Kansas, Missouri, Iowa, and Nebraska. Completing such an arduous migration, nonstop, was an impossible task, so the enormous flock often alighted to refuel. The prairies of the Midwest were favored refueling stops, and the birds used their long bills to probe the soil for insect eggs, larvae, and pupae. Interestingly, it is thought that these refueling stops were heavily dependent on the Rocky Mountain locust, another extinct animal that once lived in unparalleled aggregations.

The risks of this journey were varied and grave. The North Atlantic is ravaged by storms, and each year, many of the curlews were blown off course to find themselves alone and hungry in the cold expanse of the North Atlantic. Some stragglers even found their way to Britain and the decks of Atlantic ships. It seems that the entire world population of Eskimo curlew lived and traveled as one immense flock, which, at its peak, probably numbered in the millions. There is protection in numbers, but each year, many individuals were undoubtedly picked off by predators or perished due to exhaustion. These risks were intensified massively when Europeans started to settle North America.

Because the curlew flew in such great flocks, the settlers called them prairie pigeons, recalling the enormous flocks of passenger pigeons that blotted out the sun in eastern North America. There are accounts of an Eskimo curlew flock of 1860 measuring more than 1 km long and wide. Any animal that is edible and exists in huge numbers quickly attracts the attention of hunters, and unfortunately, the curlew was both of these things. The curlew may have seemed numerous, but the enormous flock the hunters preyed on was the entire global population of this bird, and hunting quickly took its toll. During the birds' feeding stops on their long route north, the hunters would close in on the flock and, sensing danger, the birds would take to the wing, an effective defense against land predators and birds of prey but completely useless against shotguns. The birds were so tightly spaced as they left the ground that a single blast from a shotgun, with its wide spread of shot, could easily kill 15 to 20 individuals. The birds were shot in such huge numbers that countless numbers of them were simply left to rot in big piles. The rest were taken away, piled high on horse-pulled carts. Such senseless slaughter of the Eskimo curlew on its northbound journey was bad enough, but it was not long before the hunters turned their attention to the birds' breeding grounds.

During the northern summer, in anticipation of their long migration south, the birds fed on the swarms of insects that plague the tundra in the fleeting warmth, and as a result, they grew very fat. Hunters called these well-fed birds "doughbirds," and even these were not safe. The hunters would find their roosting grounds and slaughter them under the cover of darkness, using lanterns to dazzle them and sticks to club them. The fattened birds that survived took to the wing for the start of their migration, but gales would often blow them into New England, and this was the signal for every man with a gun to come out and harvest the poor animals. In the 1830s and 1840s, the birds were blown off course and ended up in Nantucket. The populace killed the birds so mercilessly that the island's supply of powder and shot ran dry, interrupting the slaughter.

Under such intense hunting pressure, the Eskimo curlew was doomed. In 1900, Paul Hoagland was hunting with his father near Clarks, Nebraska. They scared 70 Eskimo curlews into taking flight and followed them to a newly plowed field. They killed 34 of the birds with four shots. In 1911, the same man came across eight of the birds, and he killed seven of them. Reduced from an enormous flock covering an area equivalent to around 38 football fields, this sorry collection of birds was the last to be seen in Nebraska. Since 1900, 20 Eskimo curlews have been collected by ornithologists, and in 1964, the last confirmed individual of this species was shot in Barbados. Lonely individuals may still plow the old migration routes, but it is very likely this species is gone for good.

- Hunting undoubtedly had a huge effect on the Eskimo curlew, but it is also thought that agriculture played a role in its demise. Much of the fertile prairie, the curlew's refueling ground, was turned over to agriculture, and many of the insects on which the birds fed dwindled in numbers. One example is the Rocky Mountain locust, which once lived in swarms of staggering dimensions.
- Birds that live in flocks depend on strength in numbers for protection. A lone curlew would stand little or no chance of evading predators during its arduous migratory flight. If any Eskimo curlews still remain, their continued survival will be fraught with danger and uncertainty.

Further Reading: Johnsgard, P. A. "Where Have All the Curlews Gone?" *Natural History* 89 (1980): 30–33.

CARIBBEAN MONK SEAL

Caribbean Monk Seal—Habitat loss, persecution, and competition with humans for food forced the Caribbean monk seal into extinction. (Phil Miller)

Scientific name: *Monachus tropicalis*
Scientific classification:
    Phylum: Chordata
    Class: Mammalia
    Order: Carnivora
    Family: Phocidae
When did it become extinct? The last reliable record of this species is from 1952.
Where did it live? As its name suggests, the Caribbean monk seal was native to the Caribbean region from the southeastern United States to northern South America, including tropical waters in the Florida Keys, Bahamas, and Greater and Lesser Antilles, and islets off the Yucatan Peninsula and the coast of southern Central American.

Seals, with their thick blubber, are well adapted to the chilly waters of the earth's poles and temperate regions, but monk seals, the only truly tropical seals, buck this trend and inhabit warm equatorial latitudes. Of the three species of monk seal, only the Mediterranean and Hawaiian monk seal are still around. The third species, the Caribbean monk seal, was last reliably sighted on Seranilla Bank, between Jamaica and Honduras, in 1952. On his Caribbean voyages in 1493, Christopher Columbus referred to the Caribbean monk seal as the sea wolf, a term historically used to describe various seal species, perhaps because of their habit of stealing fish from the nets and lines of fishermen. Today, most of our knowledge of what this animal looked like is based on a few photographs and observational records principally from the late 1800s and early 1900s, when at least a few small colonies still existed. Caribbean monk seals were not particularly big by seal standards. Adult males reached lengths of around 2.0–2.4 m and weighted 170–270 kg, while females were slightly smaller. As seals go, this seal was said to be an attractive animal, with grizzled brown fur tinged with gray on its back that faded to yellow on its underside and muzzle. Another characteristic feature of the seal was the hoodlike rolls of fat behind its head. For hauling its body out of the water, the nails on the seal's front flippers were well developed, while those on the rear flippers were simpler.

Although this species only became extinct in recent times and was captured in a few photographs, very little information was collected on its biology. As with the other seals, the Caribbean monk seal must have been an accomplished marine predator more at home in the water than out of it. Like other monk seals, it probably had a liking for small reef fish and eels as well as invertebrates such as octopi, spiny lobsters, and crabs. As for predators, the only animals in the Caribbean, other than humans, that could have dispatched a fully grown monk seal are sharks. In the water, the agility and keen senses of the adult seals would have made them difficult prey for sharks, although young seals unfamiliar with sharks were probably more vulnerable.

Like other seals, Caribbean monk seals spent a good proportion of their time in the water. The main times for spending extended periods out of the water were the molting season (when seals haul out to dry land and shed their old fur) and the breeding season. With little seasonal change in the tropics, the breeding season probably extended over several months and was therefore longer than the breeding seasons of most seals. Very little is known about the young of the Caribbean monk seal, although several pregnant females with well-developed fetuses were killed in the Triangle Keys off the north coast of the Yucatan Peninsula, indicating that they gave birth to their young between early December and late June. Newborn pups were around 1 m long and 18 kg in weight and were covered in dark fur.

What became of this Caribbean seal? The only confirmed sightings of this animal in the United States in the 1900s were sightings of a few individuals in the Dry Tortugas between 1903 and 1906 and the killing of lone individuals by fishermen in Key West in 1906 and 1922. The only other accounts of seals from the 1900s were off the Yucatan Peninsula, one of which involved the killing of 200 seals in the Triangle Keys. Evidently the species had already declined to very low numbers by the early part of the twentieth century due to relentless hunting. The Caribbean and its environs also underwent intense development toward the end of the nineteenth century and at the beginning of the twentieth century. As there are no land predators in the Caribbean, or at least none big enough to tackle a fully grown monk seal, this animal had no innate fear of humans. Apparently it was a curious and nonaggressive beast, a fact that made it easy pickings for hunters, who killed them for their meat and blubber, which was rendered down into oil. The seals may also have had to compete with humans for their food as the burgeoning tourist trade placed greater and greater pressure on the Caribbean's marine resources. As the human population increased in the Caribbean and demands for ocean products outstripped local supplies, fishermen turned to increasingly remote areas, where seals had been forced to retreat. As the seals were seen as a traditional resource and unwelcome competitors for their fish, the fishermen likely persecuted the last remaining seals for their blubber and meat or in self-serving attempts to protect their catch. With the combination of habitat loss, hunting, and competition for food, the monk seal was pushed to extinction.

+ Even though the last reliable sighting of the Caribbean monk seal was in 1952, people still report seeing this animal. Most of these sightings are reported by divers and fishermen, but it is highly likely that they are confusing the monk seal with hooded seals, which occasionally stray south from their northern range off Canada, or with California sea lions, which occasionally escape from navy training programs, traveling circuses, or captive facilities around the Caribbean.

+ The Caribbean monk seal is one of three monk seal species. The other two species, the Mediterranean monk seal and the Hawaiian monk seal, are both listed as endangered species and are declining. Mediterranean monk seals now number around 500 individuals, and Hawaiian monk seals number about 1,200. Hawaii and the Mediterranean are both densely populated tourist destinations, and demands for beachfront property exert direct pressure on the habitats of both seal species. It will take a lot of public awareness and active protection to ensure the survival of these animals.

+ The monk seals are a type of true seal, and they belong to a group of animals called the pinnipeds. The other members of this group are the eared seals (sea lions and fur seals) and the walruses. These semiaquatic mammals are thought to have evolved from a bearlike ancestor around 23 million years ago.

Further Reading: Boyd, I., and M. Stanfield. "Circumstantial Evidence for the Presence of Monk Seals in the West Indies." *Oryx* 32 (1998): 310–16; Debrot, A. "A Review of Records of the Extinct

W. Indian Monk Seal." *Marine Mammal Science* 16 (2000): 834–37; Kenyon, K. "Caribbean Monk Seal Extinct." *Journal of Mammalogy* 58 (1977): 97–98; Mignucci-Giannoni, A., and D. Odell. "Tropical and Subtropical Records of Hooded Seals Dispel the Myth of Extant Caribbean Monk Seal." *Bulletin of Marine Science* 68 (2001): 47–58.

THYLACINE

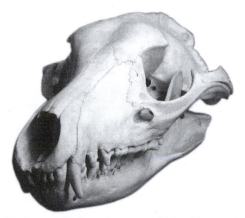

Thylacine—Only an expert would be able to tell that this skull belonged to a thylacine and not a dog. (Natural History Museum at Tring)

Thylacine—A stuffed skin of a thylacine. Note the similarity of this Australian marsupial to a dog. (Natural History Museum at Tring)

Scientific name: *Thylacinus cynocephalus*
Scientific classification:
>   Phylum: Chordata
>   Class: Mammalia
>   Order: Dasyuromorphia
>   Family: Thylacinidae

When did it become extinct? It became extinct in the year 1936, although unconfirmed sightings are still reported.

Where did it live? The thylacine was native to Australia and New Guinea, but in relatively recent times, its range was limited to Tasmania, the island off the southeastern tip of Australia.

A short, black-and-white, silent film showing an unusual doglike animal pacing up and down in a zoo enclosure is a poignant reminder of the last known thylacine, known affectionately as Benjamin. The film was shot in 1933 at Hobart Zoo in Tasmania, and three years after the film was shot, Benjamin died—some say through neglect, but whatever the cause, his demise was the end of the species.

The range of the thylacine, also inaccurately known as the Tasmanian wolf or Tasmanian tiger, once encompassed the forests of New Guinea and most of Australia, as bones and other remains testify. However, at least 40,000 years ago, humans reached these lands, and the demise of the thylacine began. When European explorers first reached this part of the world, the thylacine was restricted to the island of Tasmania, and it was already quite rare. The reason for its disappearance from the mainland is a bone of contention, but Aboriginal hunting is thought to be a factor and, much later, competition with the dingoes that first found their way to Australia via Aboriginal trading with Southeast Asian people around 4,000 years ago.

From the black-and-white film and numerous photos and accounts of the thylacine, we know exactly what it looked like and some of its behavior. In appearance, it was quite dog-like, but it was a marsupial, and like all marsupials, it had a pouch; however, unlike some other flesh-eating marsupials, the thylacine's pouch opened to the rear, and it was to this cozy pocket that the young crawled after being born, fixing themselves onto one of the four teats in its confines. As its appearance suggests, the thylacine was a predator in the same vein as other large, terrestrial, mammalian carnivores, but it had some unique features. Its jaws, operated by powerful muscles, could open very wide indeed, and its muscular, relatively rigid tail, similar to a kangaroo's, acted like a prop so the thylacine could balance quite easily on its back legs, and even hop when it needed to. We can only make educated guesses as to the animals it preyed on, but on the Australian mainland, it may have favored kangaroos and wallabies, whereas its diet on Tasmania probably consisted of just about any animal smaller than itself as well as carrion. How did the thylacine catch its prey? Again, we have to rely on accounts from the nineteenth and early twentieth century, but these vary, with some suggesting the thylacine would pursue its prey over long distances, while others report that it was an ambush predator. In Tasmania, it may have relied on both of these predatory tactics depending on the habitat in which it was hunting.

Records of the behavior of the thylacine suggest that it was active at dusk and dawn and during the night; however, this behavior may have been unnatural—a response to human persecution. During the day, thylacines built a nest of twigs and ferns in a large hollow tree or a suitable rocky crevice, and when the dusk came, they would leave these retreats in the forested hills to look for food on the open heaths.

Sadly, the thylacine's predatory nature brought it into conflict with the European settlers who started to raise livestock on the productive island of Tasmania. The killing of sheep and poultry was attributed to the thylacine, even though they were rarely seen. The authorities at the time initiated a bounty scheme in which farmers and hunters could collect a reward for the thylacines they killed. Between 1888 and 1909, this bounty was £1 per thylacine, and records show that 2,184 bounties were paid out, but it is very likely that the bounty was left unclaimed on many occasions. By the 1920s, the thylacine was very rare in the wild, and the species clung to survival as a few scattered individuals in the former strongholds of its range. Although human persecution was the final blow for this animal, it was probably also suffering from competition with introduced dogs and the diseases they carried. Benjamin was the last known thylacine, and after 50 years with no evidence of any surviving individuals, the species was declared extinct in 1986. Many people cling to the hope that a remnant population of thylacines still survives in Tasmania. Tasmania is a large, rugged, and sparsely populated island, and there is a very faint possibility that the thylacine has somehow clung to existence. The last person to photograph a living thylacine, David Fleay, searched Tasmania with a colleague, and the evidence they found suggests that the thylacine was hanging on into the 1960s. Sightings are still reported today, not only from Tasmania, but also from mainland Australia and the Indonesian portion of New Guinea. Until a live specimen of the thylacine is presented or other irrefutable evidence is declared, we have to conclude that this enigmatic species is sadly extinct.

+ The demise of the thylacine on the Australian mainland is attributed to the arrival and dispersal of Aborigines and the animals they brought with them, notably the dingo. This may only be part of the picture as the striped coat of the thylacine suggests this animal was adapted to forest. A drying of the global climate thousands of years ago may have caused Australian forest habitats to contract, and the thylacine may have been forced into areas to which it was not well adapted. This loss of habitat was compounded by the large-scale changes that followed in the wake of the first human invasion of Australia.

+ The thylacine, when compared to the wolf, is one of the best examples of convergent evolution, the phenomenon by which two unrelated animals from widely separated locations have a striking resemblance to one another because of the similar niches to which they have had to adapt. In Australasia, the thylacine filled the niche of a running predator that is occupied by canid predators in the Northern Hemisphere, and as a result, it came to look like them.

+ There are several preserved fetuses of the thylacine in museum collections around the world, and scientists had suggested that it would be possible to bring the thylacine back from extinction using the DNA from these specimens and the technology of cloning. DNA was extracted from these specimens, but it was badly degraded, and therefore cloning would have been impossible.

Further Reading: Bailey, C. *Tiger Tales: Stories of the Tasmanian Tiger.* Sydney: HarperCollins, 2001; Paddle, R. *The Last Tasmanian Tiger: The History and Extinction of the Thylacine.* Cambridge: Cambridge University Press, 2000; Guiler, E. *Thylacine: The Tragedy of the Tasmanian Tiger.* Oxford: Oxford University Press, 1985.

CAROLINA PARAKEET

Scientific name: *Conuropsis carolinensis*
Scientific classification:
    Phylum: Chordata
    Class: Aves
    Order: Psittaciformes
    Family: Psittacidae
When did it become extinct? The last Carolina parakeet is thought to have died in 1918.
Where did it live? This parakeet was a wide-ranging inhabitant of the United States.
    The two subspecies of this bird ranged from central Texas to Colorado and southern Wisconsin, across to the District of Columbia and the western side of the Appalachian Mountains, and throughout the drainage basin of the Mississippi and Missouri rivers.

Few animals have fascinated humanity for as long as the parrots and their relatives. Indigenous people in the tropics and people from Western societies alike covet these birds,

Carolina Parakeet—Stuffed skins, like this one, and bones are all that remain of the Carolina parakeet. (Natural History Museum at Tring)

not only for their beautiful appearance, but also for their playfulness and the ability of some species to mimic the human voice. The inherent beauty and charm of these birds makes it hard to understand why humans would willingly seek to wipe them out, but this is exactly what has happened on a number of occasions.

One of the most tragic examples of how humans have actively exterminated one of these interesting birds is the tale of the Carolina parakeet, a beautiful bird and the only native parrot of the United States. Around 30 cm long and 250 g in weight, this colorful bird was very common in the eastern deciduous forests of the United States, and especially in the dense woodland skirting the many great rivers of this region. The birds normally lived in small groups, although larger flocks would gather in the presence of abundant food, and it was not unusual to see 200 to 300 birds in a brilliant, raucous gathering. Like so many other parrots, the Carolina parakeet was a monogamous, long-lived species that brooded two white eggs in the cavities of deciduous trees. During most of the day, the Carolina parakeet would roost in the highest branches, and it was only in the morning and evening that the small flocks would take to the wing in search of food and water. Like other parrots, it could use its strong bill to crack open seeds and nuts to get at their nutritious contents.

The productive lands of North America suited the Carolina parakeet, and for hundreds of thousands of years, this bird brought a riot of color to the deciduous forests of this continent. Even when the first humans to colonize North America encroached on the woodlands of the Carolina parakeet, it continued to thrive. The turning point in the survival of this species came with the arrival of Europeans. The ways of the Europeans were very different to the ways of the American Indians, and they cleared large areas of forest to make way for agriculture. The Carolina parakeet was not only dependent on the forests for roosting and nesting places, but also for food. Initially, the loss of habitat did not affect the parakeet too badly as it adapted to feed on the seeds of the European's crops, including apple, peach, mulberry, pecan, grape, dogwood, and various grains. This adaptability brought the parakeet into conflict with farmers, who saw the colorful bird as no more than a troublesome pest. The slaughter of the Carolina parakeet began, and from that point on, it was doomed. Farmers would seek out the small flocks and kill one or two birds to trigger an interesting behavior that was to seal the parakeet's fate: Hearing the gunshots, the birds would take to the wing but would quickly return to their fallen flock mates, hovering and swooping over the lifeless bodies. The significance of this behavior is unknown, but it was probably a way of intimidating and confusing predators in the hope that the downed bird was only injured, thus giving it time to escape. This was probably a very successful strategy against predatory mammals and birds, but a man armed with a gun was a very different opponent. As the rest of the flock attended the bodies of the fallen, the hunter was able to pick off more of the unfortunate birds, and it was not unusual for an entire flock to be wiped out in this way.

The years passed, and the Carolina parakeet lost more and more habitat and suffered the continued persecution of ignorant humans. To make matters worse, thousands of the birds were captured for the pet trade, and thousands more were killed to supply the hat trade with colorful feathers for the latest in fashionable ladies' head wear. The senseless slaughter and collection continued, and by the 1880s, it was very clear that the Carolina parakeet was very rare. In 1913, the last Carolina parakeet in the wild, a female, was collected near Orlando in Florida, and only four years later, the last captive individual, a male by the name of Inca, died in Cincinnati Zoo only six months after the death of his lifelong partner, Lady Jane. They had lived together in captivity for 32 years. The sad and needless extinction of this interesting bird mirrors the demise of the passenger pigeon, and ironically, both species met their end in a small cage in the same zoo, poignant reminders of human ignorance, greed, and disregard for the other species with which we share this planet.

- Sightings of the Carolina parakeet were reported in the 1920s and 1930s, but it is very likely that these were misidentifications of other species that had escaped from captivity.
- Parrots, as a group, are among the most threatened of all birds. There are around 350 species of these fascinating animals, and no less than 130 of these are considered to be threatened or endangered. Unless humans can control the systematic and pathological destruction of the world's most biodiverse areas, the future looks very bleak for these birds as well as countless other species.
- Habitat destruction is not the only threat facing these birds. Thousand of parrots are collected from the wild every year to feed the ever-growing pet trade—a multi-million-dollar industry. To give you an idea of the scale of the problem, around 2 million birds are imported into the European Union every year, many illegally, and hundreds of thousands of these are parrots.

Further Reading: Poole, A., and F. Gill, eds. *The Birds of North America 667*. Philadelphia: Birds of North America, 2002.

PASSENGER PIGEON

Passenger Pigeon—Once the most numerous bird on the planet, the graceful passenger pigeon was a very fast flyer. (Natural History Museum at Tring)

Scientific name: *Ectopistes migratorius*
Scientific classification:
    Phylum: Chordata
    Class: Aves
    Order: Columbiformes
    Family: Columbidae
When did it become extinct? The last known passenger pigeon died on September 1, 1914, in Cincinnati Zoo.
Where did it live? The passenger pigeon was a native of North America, but during their winter migrations, the birds headed south, with some reaching as far as Mexico and Cuba.

In the late nineteenth century, anybody who suggested that the passenger pigeon was in danger of imminent extinction would have been branded a fool. The passenger pigeon existed in such colossal numbers that it is astonishing that it is no longer with us. The species was so numerous that there are many accounts of the bird itself and the enormous flocks in which it collected. Estimates for the total number of passenger pigeons in North America go as high as 9 billion individuals. If these estimates are anywhere near the true number, then the passenger pigeon was undoubtedly one of the most numerous bird species that has ever lived. This enormous population was not evenly spread, but was concentrated in gigantic flocks so large that observers could not see the end of them and so dense that they blocked out the sun. Some records report flocks more than 1.6 km wide and 500 km long—a fluttering expanse of hundreds of millions of passenger pigeons. We can only imagine what one of these flocks looked like, but we can be sure that it was quite a spectacle.

Apart from its propensity for forming huge flocks, the passenger pigeon was quite similar in appearance to a domestic pigeon, although it was considerably more graceful, with a slender body and long tail. Most pigeons are built for speed, but the passenger pigeon was a real racer. Its tapering wings, powerful breast muscles, and slender body gave it a real turn of speed. There is anecdotal evidence that these birds could reach speeds of 160 km per hour, although they usually flew at 100 km per hour. The aerial abilities of the passenger pigeon came in very handy as it was a migratory species. As the summer arrived in the northern latitudes, the birds would leave their wintering grounds in southern North America and head for the lush forests of the United States and Canada, although their aggregations appeared to be particularly dense on the eastern seaboard. They came to these immense forests (only remnants of which remain today) to raise young on a diet of tree seeds (mast), forming huge nesting colonies in the tall trees. As with most pigeons, the nest of the passenger pigeon was a rudimentary affair of twigs that served as a platform for a single egg. The parent birds nourished their hatchling on crop milk, the cheeselike substance secreted from the animals' crops that is unique to pigeons.

This cycle of migration had probably been going on for hundreds of thousands, if not millions, of years, but all was about to come to an end as Europeans first arrived in the Americas. Their arrival signaled the end for the passenger pigeon, and many more species besides. Europeans, in their attempts to settle these new lands, brought with them new ways and means of growing food. The forests were hacked down to make way for these crops, and the passenger pigeons were quick to exploit this new source of food. Settlers first killed the passenger pigeons to protect their crops, but they soon realized that these birds were a massive source of nutritious food, and the slaughter began in earnest. The adult birds were normally preyed on when they were nesting. Trappers equipped with nets constructed smoky fires beneath the nesting trees to force the adults into taking flight. Trees with lots of nests were cut down, enabling trappers to get their hands on the young pigeons. The slaughter was senseless and wasteful, with often only the feathers of the birds being taken to be used as stuffing. Of course, the birds were valued as cheap food, and millions of birds were taken by train to the big cities on the East Coast of the United States. It has been said that during the end of the eighteenth century and for much of the nineteenth century, servants and slaves in these big cities may have eaten precious little animal protein apart from passenger pigeon meat. For several decades, passenger pigeons ready for the oven could be bought for as little as three pennies.

By 1896, only 250,000 passenger pigeons remained, grouped together in a single flock, and in the spring of that year, a group of well-organized hunters set out to find them. Find them they did, and they killed all but 5,000 of them. Only three years later, the last birds in the wild were shot. Once the most numerous bird on the whole planet, the passenger pigeon had been wiped out in a little more than 100 years.

+ It is thought that the passenger pigeon's breeding and nesting success was dependent on there being huge numbers of individuals. Habitat destruction and hunting led to the collapse of the populations past this threshold. With their flocks in tatters and continual nesting disruption, it was not long before the population fell below recoverable levels. Scientists have also suggested that the dwindling populations of passenger pigeons could have been forced over the edge by an introduced viral infection known as Newcastle disease.

+ The nesting colonies of passenger pigeons were huge, covering an area of up to 2,200 km², which is considerably bigger than the area of Jacksonville in Florida.

+ Passenger pigeons were used to feed pigs and were processed to make oil and fertilizer. Although the adult birds were eaten in their millions, the young pigeons, known as squabs, were said to be delicious.

+ The term *stool pigeon* originates from the practice used by hunters to kill large numbers of passenger pigeons. A single bird was captured and its eyes were sewn shut with thread before it was attached to a circular stool that could be held aloft on the end of a stick. The stool would be dropped and the pigeon would flutter its wings as it attempted to land. Other pigeons flying overhead would see one of their number apparently alighting, and they, too, would land in the hope of finding food, allowing the hunters to snare them with nets.

+ Large numbers of skins and preserved specimens of passenger pigeons found their way into private collections, with at least 1,500 preserved specimens held around the world.

+ It has been suggested that before Europeans arrived and settled in North America, the populations of the passenger pigeon were held in check by Amerindian hunting. As the tribes of these people dwindled, so did their influence on the animals and plants of the eastern United States, and populations of animals like the passenger pigeon experienced explosive growth.

+ The hunting of the passenger pigeon was so intense that in 1878, a single hunter shipped more than 3 million birds to the big cities of the eastern United States. Nets and traps caught vast numbers of birds, and a variety of shotguns were used by professional hunters, marksmen, and trapshooters.

Further Reading: "A Passing in Cincinnati—September 1, 1914." In *Historical Vignettes 1776–1976*, Washington, DC: U.S. Department of the Interior, 1976; Halliday, T. "The Extinction of the Passenger Pigeon *Ectopistes migratorius* and Its Relevance to Contemporary Conservation." *Biological Conservation* 17 (1980): 157–62.

⚲ Extinction Insight: The Lottery of Fossilization

Many extinct animals are only known from bones, yet it is an often overlooked fact that the odds are stacked squarely against the remains of an animal surviving at all. It has been estimated that only one animal in billions will become a fossil. Of the billions of animals that have ever lived, only a tiny fraction have left durable remains. The dinosaurs, although considerably older than the animals mentioned in this book, are a perfect example of just how rare fossilization is. The dinosaurs are a very well studied group of fossil animals, yet in the 183 years since the first dinosaur was described, 330 species have been named. We'll never know for sure how many species of dinosaur have walked the earth, but it must have been many, many times more than 330.

The rarity of fossilization is not surprising when you consider the fate of an animal after it has died. If an animal dies in the wild, its carcass is rapidly dismembered; some bones may be cracked open, and what remains will be at the mercy of the elements. On the surface, they'll be subjected to the slightly acidic bite of rainwater, the erosive power of the wind, and the fierce rays of the sun. Being underground may afford some protection, but acidic solutions percolate through the soil, and there are countless bacteria to digest the nutrients left in the bone. In the vast majority of cases, the bones of the long-dead animal are worn away to dust and nothing remains to show it once lived. Preservation also depends on where the animal lived. If it was a denizen of warm, humid forests, the chances of preservation are even slimmer. Forests abound with scavenging animals and bacteria, and if the bones manage to find their way into the ground, the acids produced by decomposing plant matter rapidly dissolve them away to nothing.

The Lottery of Fossilization—The paleontologist Grayon E. Meade proudly poses with some of the numerous scimitar cat remains discovered in Freisenhahn Cave, Texas, during the summer of 1949. These cats' remains were buried by sediment and the cave was sealed by natural processes. They lay undisturbed for thousands of years until paleontologists discovered them during excavations. (Texas Natural Science Center)

In the rare event of a bone surviving, or even more remotely, an entire skeleton surviving the rigors of scavengers, the elements, and bacteria, something rather unique must happen. The remains must be buried quickly after an animal dies, perhaps by a freak landslide or a fall of volcanic ash, in sticky asphalt, or in a bog. With the remains well buried and protected, the process of fossilization can begin. Water percolating through the sediment or soil in which the bone lies carries silica and other materials into the pores in the bones, strengthening them and giving them the appearance of stone. Many of the animals in this book did not die long enough ago for their bones to have become completely mineralized, while others died in the wrong place for fossilization to occur. A perfect example of the latter are the remains that have been found in the dry caves of the Nullarbor Plain, Australia (see the "Extinction Insight" on the Nullarbor Plain Caves in chapter 6). The animals that died in these natural pitfall traps never got buried, and their bones lay on the floor of the cave for tens of thousands of years before being seen by human eyes for the first time. The remains of these extinct Australian animals were still just bone, albeit very delicate, as no water had ever percolated through them to leave any strengthening minerals. Similarly, the remains of so-called Flores man, recovered from Ling Bua Cave in Indonesia, had not undergone any mineralization and were on the verge of decomposing altogether.

When we think of the remains of long-dead animals, we normally think of digging around in rock to find fragments of the living animal. Although this is often the case, animal remains are preserved in other ways, some of which are astounding. In some places in Siberia and Alaska, whole animals, such as mammoths, were frozen so quickly and later buried that they are almost perfectly preserved in flesh and bone, and today they provide us with the best glimpse we have of what these ice age animals were like. In very dry places, a dead animal can become mummified. Some ground sloths have been preserved in this way, and even though the vast majority of their soft tissues have been eaten by insects and other small animals, fragments of skin and hair, thousands of years old, remain. Some animals met their end in peat bogs, and these deep beds of slowly decomposing plant matter are excellent for preservation of animal bones and even soft tissues. Tar pits, like peat bogs, keep oxygen away from the remains of dead animals, and the bones that come to lie in these pools of ooze are remarkably well preserved.

The fossil record may be very fragmentary, but it is continually being added to. With every passing day, new fossils are revealed as the action of water, wind, and ice erodes the surface of the earth.

Earth's secrets are revealed to us slowly, and as scientists continually explore the far corners of our planet, searching for the remains of animals, they will add to our knowledge of what the earth was like and how it is changing. With every passing year, new species are added to the list of animals that were. Who knows what remarkable creatures will be found buried in sediment or frozen in permafrost in the future? The remains of some unknown animals will come to light only to be eroded away by the very forces that revealed them, and the only evidence of their existence will be lost forever.

2

FEWER THAN 200 YEARS AGO

ROCKY MOUNTAIN LOCUST

Rocky Mountain Locust—The Rocky Mountain locust formed enormous swarms, possibly the largest known aggregations of any animal. (Phil Miller)

Scientific name: *Melanoplus spretus*
Scientific classification:
  Phylum: Arthropoda
  Class: Insecta
  Order: Orthoptera
  Family: Acrididae
When did it become extinct? The last sighting of this insect was in 1902.
Where did it live? The native range of this insect was the eastern slopes of the Rocky Mountains, extending from the southern forests of British Columbia through Montana, Wyoming, Idaho, and the western parts of the Dakotas. In some years, the species was able to extend its range to take in one-third of Manitoba, the Dakotas, Minnesota, Kansas, Oklahoma, Missouri, the western half of Nebraska, and the northeastern part of Colorado.

In the late nineteenth century, much of the United States was a frontier where people sought to realize their American dream, and many of them headed to the vast prairies of this continent. The term *prairie* conjures up images of beautiful, undulating plains stretching as far as the eye can see, yet this image is not altogether accurate. In the winter, these plains get bitterly cold, and in the summer, they are blistering hot. Add to this an almost perpetual wind, and what you get is an unforgiving environment. As if these tough conditions weren't enough for the settlers, they were also confronted with an insect that amassed in swarms of a gargantuan nature.

The Rocky Mountain locust was small by typical locust standards, with an adult body length of 20 to 35 mm, long wings that extended past the end of the abdomen, and the enlarged back legs common to most grasshoppers. What this insect lacked in individual size it more than made up for in the size of its aggregations. Locusts, for much of the time, live their lives in the same way as most other grasshoppers—going about their business without being much of a nuisance to anyone—but occasionally, their populations may become very dense, and this triggers a dramatic change. The locusts change color, their wings grow, and they start to amass in swarms.

The swarms formed by the Rocky Mountain locust were incredible and probably represent some of the biggest aggregations of any land animal that has ever existed. A swarm observed in Nebraska during the summer of 1874 was of staggering proportions. Dr. A. L. Child of the U.S. Signal Corps was charged with assessing just how big this swarm was, and to get an idea, he measured the speed of the locusts as they were flying past and then telegraphed surrounding towns to get an idea of its extent. The swarm was estimated to be about 2,900 km long and 180 km wide. Observers in the Nebraskan towns over which this swarm passed reported that the gigantic cloud of insects obscured the sun and took five days to pass overhead. This begs the question of how many locusts there were in this enormous swarm. Estimates are as close as we'll ever get, but it has been calculated that there must have been around 12 trillion insects in this aggregation. All these fluttering insects weighed somewhere in the region of 27 million tonnes, and if the desert locust of the Old World is anything to go by, then this swarm may have eaten its own weight in food every day just to sustain itself. Luckily, the Rocky Mountain locust was not a fussy eater—it would nibble a huge range of plants, and in the absence of foliage, it would munch bark, leather, laundry, dead animals, and even the wool off a sheep's back. As can be imagined, the multitude of mandibles left a trail of devastation, and between 1873 and 1877, the vast swarms of insects caused massive crop damage in Nebraska, Colorado, and some other states, estimated at around $200 million.

Around 30 years after these immense swarms left a trail of devastation in their wake, the Rocky Mountain locust mysteriously vanished. The reason behind the extinction of this insect has been speculated on for some time. Some experts have suggested that the species never became extinct and that the locust was actually the swarming phase of a species that can still be found today, a theory that has been shown to be incorrect. The likely explanation for the disappearance of this insect is that outside of its swarming periods, the locust retreated to the sheltered valleys of Wyoming and Montana, where the females laid their eggs in the fertile soil. These very same valleys attracted the attention of settlers, who saw their potential for agricultural endeavors, and with their horses and their plows, they turned the soil over and grazed their livestock on the nutritious grass. These actions destroyed the eggs and developing young of the insect, and around three decades after its swarms blotted out the sun, the Rocky Mountain locust was gone forever.

- The swarming of grasshopper species, such as the Rocky Mountain locust, is thought to be a survival mechanism that allows the insects to disperse into new habitats when things get a little cramped during periods of worsening environmental conditions that concentrate the nymphs into ever shrinking areas. In their normal or solitary phase, the grasshoppers are very sensitive to the presence and proximity of others of their kind. When things start to get a bit too cozy, the insects switch from intolerance to attraction, forming so-called bands of nymphs. The locusts take on the appearance of the swarming insect and fly off in search of more space and food.

- Settlers in the native range of the locust also killed huge numbers of beavers and widened streambeds, both of which led to increased flooding and the death of locust eggs and young in the ground. These settlers also planted alfalfa over huge swathes of ground, a plant that the locust was not fond of eating. It has also been suggested that bird species from the eastern United States followed the settlers along corridors of cottonwood, preying on huge numbers of insects, including the locust.
- Female Rocky Mountain locusts used a pair of tough valves at end of their abdomens to excavate a tunnel and deposit their eggs below the surface of the soil, where they would be out of the sight of most predators. For added protection, the eggs were cocooned in a hardened foam egg sac with the appearance of a stale marshmallow.
- Some of the glaciers of the Rocky Mountains are known as grasshopper glaciers as large numbers of Rocky Mountain locusts from the swarms were driven by winds high up into the mountains, where they perished on the glaciers, only to be covered by subsequent layers of snow and ice. As these glaciers thaw, they reveal the mummified remains of these insects.
- Although the Rocky Mountain locust was very numerous, surprisingly few specimens are to be found in collections. Entomologists at the time saw little point in collecting such numerous animals, as it was inconceivable to them that an insect forming such vast swarms could ever become extinct.

Further Reading: Chapco, W., and G. Litzenberger. "A DNA Investigation into the Mysterious Disappearance of the Rocky Mountain Grasshopper, Mega-Pest of the 1800s." *Molecular Phylogenetics and Evolution* 30 (2004): 810–14; Samways, M.J., and J.A. Lockwood. "Orthoptera Conservation: Pests and Paradoxes." *Journal of Insect Conservation* 2 (1998): 143–49; Lockwood, J.A., and L.D. DeBrey. "A Solution for the Sudden and Unexplained Extinction of the Rocky Mountain Grasshopper (Orthoptera: Acrididae)." *Environmental Entomology* 19 (1990): 1194–1205; Lockwood, J.A. "Voices from the Past: What We Can Learn from the Rocky Mountain Locust." *American Entomologist* 47 (2001): 208–15; Lockwood, J.A. *Locust: The Devastating Rise and Mysterious Disappearance of the Insect That Shaped the American Frontier.* New York: Basic Books, 2004.

PIG-FOOTED BANDICOOT

Scientific name: *Chaeropus ecaudatus*
Scientific classification:
    Phylum: Chordata
    Class: Mammalia
    Order: Peramelemorphia
    Family: Chaeropodidae
When did it become extinct? The last verifiable specimen was collected in 1901, but it probably survived in remote areas for far longer, possibly until the 1950s.
Where did it live? This marsupial was known only from the plains of inland Australia.

Pig-Footed Bandicoot—The pig-footed bandicoot was a small, fleet-footed marsupial from the plains of Australia. (Phil Miller)

Australia was once home to a unique collection of beasts, including giant marsupials and fearsome reptiles. However, scurrying around the big feet of this megafauna were a huge number of small marsupials that evolved to fill most of the ecological niches occupied by placental mammals in other parts of the world. There were rabbitlike marsupials, tiny mouselike animals, even a marsupial equivalent of a mole, to name but a few. Some of these animals can still be found today, but many ended up going the same way as the other long gone denizens of Australia.

The pig-footed bandicoot was one of these animals. For millions of years, this odd little marsupial, which was no bigger than a kitten, lived throughout Australia, but in recent times, it became restricted to the arid and semiarid inland plains. This bandicoot, with its rabbit ears, was probably a familiar sight to the Australian Aborigines as it hopped and bounded around the plains.

Perhaps the oddest thing about this marsupial was the four spindly legs that supported its plump little body. It is from the animal's feet that we get its common name. On its forefeet, there were only two functional toes with hooflike nails, remarkably similar to the feet of a pig, but in miniature. The hind limbs were also highly modified as the second and third toe were fused together, and only the fourth toe, which ended in a nail like a tiny horse's hoof, was used in locomotion. With such highly modified limbs, the pig-footed bandicoot was undoubtedly a running animal, and the gait it used depended on how fast it was moving. When it was skulking around looking for food, the pig foot moved in a series of bunny hops—taking its weight on its forelimbs and pulling its back legs along. When it chose to up the pace, the hind limbs were moved alternately and, according to Aborigines, when it really wanted to move, it stretched out and took to a smooth gallop. Not only was the pig foot quick, but it also had a lot of stamina and could run at full speed for long periods of time.

Apart from being very fleet of foot, the pig foot was also said to be more dependent on plant food than the other types of bandicoot, which are generally insectivorous marsupials. In the wild, they subsisted on grass seeds, but in captivity, they ate a range of food, including lettuce, bulbs, and grasshoppers. It is said that during the hottest part of the day, they would seek refuge from the sun's rays in a grass nest, only venturing out to seek food and mates in the early evening. If the other bandicoots are anything to go by, the pig foot must have had a very short gestation. Baby bandicoots spend only about 12 days in their mother's womb—the shortest time for any mammal—and they are also unique for being attached to their mother by a placentalike organ. The pig foot's short gestation probably ended in a very short birth—which, for living bandicoots, is around 10 minutes. The tiny babies crept to their mother's rear-facing pouch, and although there were eight teats in this furry pocket, there were no more than four babies in each litter. After the young had outgrown the pouch, the female left them in a grass nest until they were ready to follow her on forays for food in the warmth of the evening sun.

What happened to the pig foot? The last known definite specimen was collected in 1901, and even long before this date, it was never considered to be a common species. We do know that it was hunted by Australian Aborigines for its meat, which was regarded as a delicacy, and its tail brush, which was sometimes worn as a decoration. The extinction of some of Australia's other native animals has been blamed on Aborigines, but the pig-footed bandicoot coexisted with the Aborigines for thousands of years. The decline and extinction of this unique marsupial coincides with the spread of Europeans through Australia. For thousands of years, Aborigines practiced brush burning to clear land and encourage new plant growth. Many species of smaller marsupial profited from this because of the food it provided, not only in terms of fresh plant matter, but also in terms of the smaller animals that were forced out of hiding by the smoke and flames. With the arrival of Europeans, all this changed, as the Aborigines themselves were pushed toward extinction. The way the Aborigines managed the land ended, and any native animals that had previously benefited were faced with some tough times. As the Europeans swept aside the old Aboriginal ways, they replaced them with their own methods of taming the harsh land. They brought modern agriculture and a menagerie of domestic animals, including dogs, cats, foxes, sheep, goats, and cattle. To a seasoned predator, such as a cat or fox, the pig-footed bandicoot must have been a delightful morsel; however, hunting by introduced species was probably only a minor factor in their extinction. Agriculture probably had the greatest effect on this species. Herds of sheep, goats, and cattle grazed the delicate plains of inland Australia, lands that simply could not tolerate the intensive chomping of countless mouths, not to mention the hordes of hooves, which churned the ground into a dust bowl. Not long after Europeans first settled Australia, the pig-footed bandicoot joined the long roll call of extinct marsupials.

+ Although the last verifiable pig-footed bandicoot was collected in 1901, interviews with Aborigines suggest that it may have survived until the 1950s in some parts of the remote interior. As this animal is so small and shy, there is an outside chance that it survives today in some forgotten corner of inland Australia.
+ The Australian zoologist Gerard Krefft sought the help of Aborigines to help him find some specimens of the pig-footed bandicoot. The picture he showed them was a pig-footed bandicoot, but it lacked a tail, and so after several false starts, where they brought him other bandicoot species, he was delighted to see a pair of pig foots. He

kept these animals for some time and recorded his observations, but when he realized his supplies were running a bit low, he ate them both. This is not the only time that science has lost out to the appetite of some famished pioneer.

Further Reading: Burbidge, A., K. Johnson, P.J. Fuller, and R.I. Southgate. "Aboriginal Knowledge of the Mammals of the Central Deserts of Australia." *Australian Wildlife Research* 15 (1988): 9–39.

QUELILI

Quelili—Collectors were remorseless in their pursuit of the quelili, and the last examples of this Guadalupe caracara were seen in 1901. (Renata Cunha)

Scientific name: *Caracara lutosa*
Scientific classification:
　　Phylum: Chordata
　　Class: Aves
　　Order: Falconiformes
　　Family: Falconidae
When did it become extinct? The last reliable sighting of this bird was in 1901.
Where did it live? This bird of prey was found only on the island of Guadalupe.

Two hundred and forty miles off the northwest coast of Mexico lies the island of Guadalupe, a small volcanic island, 35 km long and about 9 km at its widest point. Even though it is barely a speck in the vastness of the Pacific Ocean, Guadalupe was once home to a number of animals that were found nowhere else. One of the most famous Guadalupe residents was the quelili. This bird of prey was very closely related to the caracaras of Central and South America, and perhaps the ancestors of the quelili found themselves on the remote, rocky outpost of Guadalupe after being blown from the mainland during a storm.

The caracaras are all meat eaters, but they don't have the hunting prowess of eagles or falcons. They are quite feeble flyers and are unable to swoop on their prey from a great height. Instead, they prefer to catch and eat small prey that can be easily overpowered, and they often resort to scavenging. The English-speaking inhabitants of Guadalupe called the quelili the "eagle," but like the other caracaras, the quelili was no formidable aerial hunter. It apparently fed on small birds, mice, shellfish, worms, insects, and carrion when the opportunity arose.

There are a few accounts of how the living quelili behaved. Its broad wings were suited to loping flight quite close to the ground, and like the other caracara species, it may have been equally at home on the ground, stalking among the low vegetation on its long legs. Small flocks of these birds were often seen in flight, but it is unclear if there was an ordered social structure. Living caracaras are normally solitary, but they will tolerate each other around a carcass, albeit with bouts of noisy quarreling. Perhaps the quelili was a little friendlier to others of its kind. They were known to communicate with complex displays, one of which involved the bird extending its neck to full length and then arching backward until its head almost touched its back (the crested caracara displays in the same way). Unfortunately, the significance of these displays is now lost, but perhaps it was the way that one quelili asserted dominance over another.

The quelili was probably the dominant predatory land animal on Guadalupe for tens of thousands of years, but due to its position in the food chain and the small size of its island home, it would never have been very common. An island like Guadalupe could have never supported more than a couple hundred quelili, but in the narrow geological window in which it lived, this bird was a successful scavenger and predator.

This success continued up until the early eighteenth century, at which time humans appeared on the scene. The first humans to make any real difference to the ecology of Guadalupe were whalers and hunters, who came to catch and kill sea otters, fur seals, and elephant seals. On their ships, they carried goats as a source of meat and milk, and as a way of caching supplies on their hunting routes, they left some goats on Guadalupe. The idea was that the goats would survive and the whalers could pick up some fresh meat and milk the next time they were passing. Not only did the goats survive, but they bred in profusion, and before long, there were thousands of them running riot over the once virginal land. Goats in the wrong place can be devastating, as any gardener will attest. They eat anything and everything, and the numerous unique plants that covered Guadalupe were stripped away by thousands of hungry mouths. This in itself was not the nail in the coffin of the quelili, but the huge herds of goats soon attracted people. Some came to herd the goats and others came to hunt them, and herder and hunter alike both considered the quelili to be a meddlesome foe that would kill and eat goat kids whenever the opportunity arose. It is very unlikely that the quelili could have captured and killed a healthy goat kid, but it was probably partial to the flesh of a goat carcass. Goatherders may have seen a group of quelili tearing at the carcass of a dead goat kid and presumed the birds were responsible for its death.

By the nineteenth century, the quelili was goat enemy number one and it was hunted mercilessly. By the 1860s, rifles and poison had pushed it to the brink of extinction. As if angry goatherders were not bad enough, the quelili soon found itself pitted against an even more relentless foe: the ornithological collector. The age of discovery gripped the educated world, and the race to collect and catalogue the world's treasures was well and truly on. Rarities are coveted by collectors, and institutions and wealthy individuals soon got wind of the disappearing Guadalupe bird fauna, including the quelili. Back in the nineteenth century, the word *conservation* didn't really exist, and the collectors systematically exterminated the quelili; the skins were sold to the highest bidder.

Amazingly, one small group of quelilis survived this onslaught, but these were accounted for by Rollo Beck, an ornithologist and collector who landed on the island on December 1, 1900. No sooner had he landed on the island than he saw a flock of 11 quelili heading straight for him. In the mistaken belief that the bird was still common, he shot all but two of the flock, and in doing so, Rollo Beck consigned the quelili to extinction.

+ Guadalupe was once home to an array of unique plants and animals, but we know only a fraction of what species the island once supported. At least six species and subspecies of bird have become extinct since humans first colonized the island.
+ Guadalupe was covered in distinct vegetation types, ranging from areas of succulent herbs to forests of endemic cypress. Today, almost all of this has disappeared and most of the vegetation is little more than a few centimeters tall, all thanks to the tireless mouths of the introduced goats.
+ Guadalupe is governed by Mexico, and even though the island has been a protected reserve since 1928, only recently has anything been done to restore the habitats on the island. In 2005, a scheme was initiated to remove the goats from the island, and it is hoped that once these destructive herbivores are gone, the island's vegetation will regenerate naturally.
+ Caracara bones from the Rancho La Brea asphalt deposits, approximately 40,000 years old, are supposedly very similar to quelili bones, and as California is so close to Guadalupe, there is a good chance that this is where the ancestors of the quelili originated.

Further Reading: Abbott, C.G. "Closing History of the Guadalupe Caracara." *The Condor* 35 (1933): 10–14.

STEPHENS ISLAND WREN

Scientific name: *Xenicus lyalli*
Scientific classification:
> Phylum: Chordata
> Class: Aves
> Order: Passeriformes
> Family: Acanthisittidae

When did it become extinct? This small bird is thought to have become extinct in 1894.
Where did it live? The wren was found only on Stephens Island, New Zealand.

Stephens Island Wren—A cat and a lighthouse keeper almost certainly drove the tiny, flightless Stephens Island wren into extinction. (Renata Cunha)

Rising to heights of around 300 m, Stephens Island looms off the northernmost tip of Marlborough Sound on South Island of New Zealand. The island is tiny (2.6 km^2), but it is a refuge for many animals that have disappeared from the mainland since the arrival of Polynesians.

On this prominent lump of rock, there once lived a small bird known as the Stephens Island wren. This bird was unrelated to the familiar wrens of the Northern Hemisphere and actually belonged to a small group of perching birds endemic to New Zealand. The remains of this small bird have been found at various sites throughout the main islands of New Zealand, and it seems that Stephens Island was the last refuge for this bird following the arrival of humans and the animals they brought with them. One animal in particular, the Polynesian rat, wreaked havoc among the populations of New Zealand's small endemic birds. Stephens Island served as a refuge for the wren for hundreds of thousands of years, and even after the Polynesians and their animals wiped out these small birds on the mainland, the population on Stephens Island was safe—until the arrival of Europeans.

The British commandeered New Zealand as an extension of their growing empire, and in their learned opinion, what Stephens Island needed more than anything was a lighthouse to warn ships away from the rocks. In June 1879, a track to the proposed site for the lighthouse was cleared, and five years later, the lighthouse went into operation. In itself, the lighthouse was no threat to the wren, but in those days, lighthouses were operated by people, and people have pets—often, cats.

At some point in 1894, a pregnant cat was brought to the island, and it seems that no sooner had she arrived than she gave her new owners the slip and escaped. This unassuming cat probably didn't realize how special she was. No predatory land mammal had ever set foot on Stephens Island, and the animals on this forested outcrop were woefully ill prepared as they had never encountered any mammal, let alone one with the predatory proclivities of the domestic cat. In June 1894, one of the offspring of the escaped cat was apparently taken in by one of the assistant lighthouse keepers, David Lyall. Lyall had an interest in natural history, and he was intrigued by the small carcasses his young pet brought back from its forays around this previously untouched island. The carcasses were those of a tiny bird, but of a sort that Lyall had never seen. With a hunch these birds were something special, he had one sent to Walter Buller, an eminent New Zealand lawyer and ornithologist, who immediately recognized the sorry-looking carcass as an undescribed species. The bird was definitely a type of New Zealand wren, related to another small New Zealand bird, the rifleman. Unlike the rifleman, the Stephens Island bird was flightless. The larger group to which these birds belong, the perching birds (passerines), has only a couple of flightless representatives.

The only information we have on the way the Stephens Island wren lived comes from the limited observations made by Lyall. According to the only person who saw this species alive, it "ran like a mouse" and "did not fly at all." This is about the sum of the information we have on the living bird, but the structure of the bird's skeleton and plumage allows us to investigate if Lyall was correct. The skeleton of this tiny bird bears all the hallmarks of a species that had given up the power of flight, and the plumage does not appear to be up to the job of flapping flight. We can't rule out the possibility that this tiny bird ran and leapt or glided to catch aerial insects, but it would not have been capable of flapping its wings to any great effect. The great tragedy is that this tiny bird died out before we could learn anything more about it.

In 1894, Lyall brought a total of 16 to 18 specimens of Stephens Island wren to the attention of the scientific establishment. It is not clear if his cat caught all of these, but late in 1894, news of this bird had circulated in the ornithological community, and some collectors were willing to pay big money for a specimen—Lionel Walter Rothschild, the famous British collector, purchased nine specimens alone. With such a high price on the heads of these diminutive birds, can we be sure that Lyall didn't go and catch some himself to supplement his income? We'll never know, but the cats and the greed were too much for the Stephens Island wren, and before 1894 was out, the species was extinct—discovery and extinction all in the space of one year. This is pretty impressive, even by human standards of devastation.

- Of the 16 to 18 specimens collected and sold by Lyall, only 12 can be found today in museum collections around the world. This is all there is of this interesting little bird.
- The prices paid for a Stephens Island wren in 1895 are astonishing. Lyall's middleman, a man by the name of Travers, was offering two specimens for £50 each. In 1895, an average lighthouse keeper's annual salary was £140.
- Currently we don't how the ancestors of the Stephens Island wren managed to cross the 3.2 km of ocean to reach the island from the mainland. The populations of this bird on the mainland were also flightless, so it must have floated to the island on rafts of vegetation. Stephens Island is also home to one of the rarest amphibians in the world, Hamilton's frog. This animal will die if it is immersed in seawater for any length of time, so it, too, must have floated across to Stephens Island on large rafts of vegetation that were detached from riverbanks during floods and storms.

+ Today, Stephens Island is once more a safe haven for a range of endemic New Zealand animals, including the ancient tuatara and lots of weta, the giant insects that fill the ecological niche comparable to that occupied by mice and other rodents elsewhere in the world.

Further Reading: Millener, P. R. "The Only Flightless Passerine: The Stephens Island Wren (*Traversia lyalli*: Acanthisittidae)." *Notornis* 36 (1989): 280–84.

TARPAN

Tarpan—A pair of tarpan stallions fight during the breeding season. This hardy animal is widely considered to be the ancestor of most modern horses. (Renata Cunha)

Scientific name: *Equus ferus*
Scientific classification:
  Phylum: Chordata
  Class: Mammalia
  Order: Perissodactyla
  Family: Equidae
When did it become extinct? The last known pure-bred tarpan died in 1887.
Where did it live? The tarpan was native to the steppes of central Asia.

It may come as a surprise, but the domestication of the horse stands out as one of the most significant moments in human history. This seemingly insignificant event changed the way we lived forever. It enabled our ancestors to travel quickly over huge distances, and they harnessed the strength and tenacity of these animals to do tasks that previously required several men. Also, when the useful life of the horse was over, its flesh provided sustenance and its skin, bones, and sinews were turned to a multitude of uses.

What are the origins of these first domestic horses? What were they, where did they come from, and how did they live? It is widely accepted that the ancestor of the majority of modern horses was an animal known as the tarpan. This sturdy horse was only around 1.5 m at the shoulder and therefore very small compared to a modern Thoroughbred racehorse. However, what the tarpan lacked in size it more than made up for in resilience and stamina. Being an animal of the Asian steppes, it was able to survive in the very harsh conditions that sometimes sweep over these treeless plains. In the wintertime, its grayish brown coat grew long to give it added protection from the cold. In some of the more northern reaches of its range, the tarpan may even have been white. According to some of the Evenk people, ivory hunters searching for the tusks of mammoths in the deep permafrost of Siberia would often find white horses. It is possible that these could have been white tarpan that met their end in a bog, only to become entombed in ice as the earth entered another of its many glaciations.

Like other horses, the tarpan was a grazer and a herd animal. Like many other fleet-footed animals, the tarpan found protection from its predators by living in a herd. Long ago, the Asian steppe was prowled by many different predators, many of which were perfectly able to catch and subdue an animal as large as the tarpan. One by one, the tarpan's predators died out, leaving only the wolf, the occasional bear, and of course, humans. By all accounts, the tarpan was a very spirited animal and quite capable of defending itself by kicking and biting. Humans are known to have killed the tarpan by driving herds of them off cliffs, a surefire way of killing lots of them quickly.

Horses are shown in many cave paintings throughout Europe, and it is very likely that the tarpan and its relatives were simply hunted before an ancient innovator thought it would be a good idea to try to tame them. Hunting these animals on the steppes must have been very hard as horses have excellent smell and hearing and can sense the approach of danger way before they can see it. When the domestication breakthrough came, hunting was made much easier on the back of a tame tarpan, and the species began its slow, inexorable slide toward extinction. Hunting was not the main problem facing this species. As people became aware of the usefulness of the tarpan, more and more would have been taken from the wild to supplement the young that were reared from the tame individuals. The numbers of the domesticated tarpan grew, and over time, their distinctive characteristics, such as aggression and spiritedness, were filtered out in the process of selective breeding to produce a horse that was calm and cooperative. These animals were less like tarpan and more like today's horses. Unfortunately for the tarpan, it could still mate with these domesticated horses, and its unique genes were diluted. This continued until the middle of the nineteenth century, when it was realized that purebred tarpans were very rare. In 1879, the last wild tarpan was killed, but some had been taken into captivity years before and were often kept on the private estates of noblemen. These captive animals dwindled due to neglect, and the last one died in Poland in around 1887. When the tarpan became extinct, domesticated horses had found their way all over the world, as human explorers took them wherever they went.

In a vain attempt to resurrect the tarpan, the Polish government collected together a number of ponies that were considered to have tarpan characteristics. These were taken from their peasant owners and sent to forest reserves. This was a pointless exercise as the ponies they chose were a product of millennia of selective breeding and they were no more purebred tarpan than a German Shepherd dog is a purebred wolf. The same German scientists who thought it would be possible to resurrect the aurochs turned their attention to recreating the tarpan by selective breeding. This notion sorely lacked merit because no one knew or knows to this day what constitutes the tarpan on a genetic level. These attempts at selective resurrection did produce two types of horse, the Konik of Poland and the Heck of Germany, which are thought to resemble the tarpan superficially.

The story of the tarpan is an interesting one because it's not a simple case of a species being extinguished. Through our desire to produce an animal that was of use to us, we took the tarpan and molded it to our own needs, in the process producing something quite distinct. The tarpan our ancestors knew is no longer with us in a form they would recognize, but its genes are there in the cell of almost every horse.

- For a long time, scientists have been piecing together the story of horse evolution, and now they have several important parts of the puzzle. The first clear ancestor of the horse, *Hyracotherium*, evolved around 10 million years after the extinction of the dinosaurs in North America. About the size of a fox, this animal had four of its five digits in contact with the ground, and adaptations for running were already apparent, for example, long, thin legs. Over millennia, these primitive horses gradually assumed the appearance of the modern horse, with the key feature of having only one digit in contact with the ground, making them fleet-footed animals of the plains.
- Today, the only surviving truly wild horse is Przewalski's horse, a sturdy, pony-sized animal that roams the wilderness of Mongolia. Extinction almost claimed this horse, too, but captive specimens allowed a breeding and reintroduction program, which has returned small numbers of these animals to the wild.

Further Reading: Jansen, T., P. Forster, M. A. Levine, H. Oelke, M. Hurles, C. Renfrew, J. Weber, and O. Olek. "Mitochondrial DNA and the Origins of the Domestic Horse." *Proceedings of the National Academy of Sciences USA* 99 (2002): 10,905–10.

QUAGGA

Scientific name: *Equus quagga quagga*
Scientific classification:
    Phylum: Chordata
    Class: Mammalia
    Order: Perissodactyla
    Family: Equidae
When did it become extinct? The last quagga, a captive specimen, died in 1883.
Where did it live? The quagga was only found in South Africa, particularly in the Cape Province and the southern part of the Orange Free State.

Quagga—A subspecies of the plains zebra, the quagga retains some degree of striping. (Natural History Museum at Tring)

The quagga, like the dodo, is one of the more familiar animals that has gone extinct in recent times. Amazingly, this horselike animal was wiped out before anyone had figured out what it truly was. In Victorian times, it was the trend among naturalists to describe new species wherever and whenever possible, and the zebra of Africa received a good degree of attention from these early taxonomists. Zebras vary widely in size, color, and patterning, and all of these subtle differences were thought to represent subspecies and even distinct species. With the advent of molecular biology and DNA sequencing, it rapidly became clear that there was little validity in what the gentleman scholars of the previous age had proposed. Very recently, scientists managed to isolate some DNA from the mounted skins of the quagga that can be found in several museums around the world. It turned out that the quagga was very likely a subspecies of the plains zebra and not a distinct species at all.

Sometime between 120,000 and 290,000 years ago, the population of plains zebras in South Africa became isolated from the rest of their species and they started to take on a slightly different appearance. The major difference between the quagga and the plains zebra is the animals' coat. Live specimens of the quagga only had obvious stripes on their head and neck, but even the 23 specimens in the world's museums exhibit a lot of variation, with some specimens having more stripes than others. The unusual name "quagga" comes from the Hottentot name for the animal, *quahah*, in imitation of the animal's shrill cry. Aside from these details, quaggas lived like the plains zebras that can still be seen in sub-Saharan Africa today. They lived in great herds and could often be found grazing with wildebeest or hartebeest and ostriches. It has been suggested that grazing together afforded these animals greater protection from their principal enemy, the lion, thanks to a combination of their talents: the birds' eyesight, the antelopes' sense of smell, and the quaggas' acute hearing. A lion would have been hard-pressed to surprise a group of animals cooperating in this way, and it is very likely that lions caught very few healthy adult quaggas.

This defense was very effective against lions, but it was not so successful against the Boers, who were equipped with horses and guns. As the Boers moved inland, they exterminated these giant herds of ungulates, primarily for food but also for their high-quality skins. Quaggas were also captured live and put to various uses. By all accounts, the quagga was a very lively, highly strung animal, and the stallions were prone to fits of rage, so taming one of these animals must have been very interesting and practically impossible. In the early days of the Boers' settlement of South Africa, the quagga was sometimes kept as a guard horse to protect domestic livestock. Any intruder, be it a lion or a rustler, was treated to the whinnying alarm of the quagga and most probably attacked by this tenacious horse. Some quaggas also found their way to Europe, where they ended up in the big zoos. The powers that be at London Zoo thought a quagga breeding program would be an excellent idea; however, this quickly came unstuck when the lone stallion lost its temper and bashed itself to death against the wall of its enclosure. Regardless of the quagga's spirited nature, it seems there was a trend for quaggas as harness animals, and the cobbled streets of 1830s London rang out to the sounds of their cantering hooves. Just how they were coaxed into pulling a carriage full of genteel Londoners is unknown, but they were probably gelded beforehand.

The Boers, and the British before them, were quick in taming the verdant lands of South Africa, lands that abounded in game and opportunity. The native tribes of South Africa fought these invaders but were forced to abandon their prime territories. The Europeans mercilessly destroyed the abundant South African wildlife, not only for food and skins, but also for recreation and to make way for agriculture. The quagga was one of the casualties of this onslaught. In the 1840s, great herds of quaggas and other animals roamed South Africa, but only 30 years later, in 1878, the last wild quagga was shot dead. The last quagga, a female, died in Artis Magistra Zoo in Amsterdam in 1883. Today, the remnants of this South African wildlife can only be seen in national parks.

- Six subspecies of the plains zebra are recognized. Two of these, the quagga and Burchell's zebra, are extinct today, and the other subspecies have lost a lot of their habitat to human encroachment. Although their numbers have declined, zebras can still be seen in large numbers in sub-Saharan national parks.
- As with the tarpan and the aurochs, animal breeders are attempting to resurrect the quagga by selectively breeding from living zebras that have quagga characteristics. Such an exercise is quite pointless, and the resources needed for such programs would be much better spent protecting the surviving zebras.
- According to analysis of quagga DNA, this subspecies became isolated from the plains zebra sometime between 120,000 and 290,000 years ago. If correct, this is a remarkably short amount of time for the differences seen in the outward appearance of the quagga to evolve. Perhaps a population of the plains zebra was completely isolated in South Africa and started to evolve along a unique course. This is the very beginnings of speciation, the process where one species becomes two over thousands or millions of years. After less than 300,000 years, the quagga had almost lost the distinctive coat

of the plains zebra, and if it had been allowed to survive for thousands more years, it would have continued to differentiate until it was a distinct species in its own right.

Further Reading: Leonard, J.A., N. Rohland, S. Glaberman, R.C. Fleischer, A. Caccone, and M.A. Hofreiter. "Rapid Loss of Stripes: The Evolutionary History of the Extinct Quagga." *Biology Letters* 1 (2005): 291–95.

WARRAH

Warrah—The Falkland Island fox, or warrah, was the only large land mammal on the windswept archipelago in the South Atlantic. (Phil Miller)

Scientific name: *Dusicyon australis*
Scientific classification:
    Phylum: Chordata
    Class: Mammalia
    Order: Carnivora
    Family: Canidae
When did it become extinct? The last known warrah was killed in 1876.
Where did it live? This carnivore was known only from the Falkland Islands.

Remote and treeless, the Falkland Islands is a small archipelago in the South Atlantic Ocean. Ravaged by incessant winds and terrible winter storms, these islands are a very harsh environment. Although the Falklands are a welcome refuge for marine animals such as penguins, seals, and sea lions, very few land animals have managed to make a living on this stark, oceanic outpost. The only mammals known from the Falkland Islands are a small species of mouse and a mysterious dog, the warrah, which also goes by the names of "Falkland Island fox" and "Antarctic wolf."

Whether the animal was a fox or a wolf is a bone of contention among mammal experts. Contemporary accounts of the living animal as well as stuffed skins show that this carnivore had both wolf and fox characteristics. An adult warrah was about twice as big as a red fox (1.6 m long), with a large, wolfish head, but because of its short legs, it was only about 60 cm tall at the shoulder. Its tail, unlike that of a wolf, was thickly furred, and like a fox, it excavated dens in the sandy soil of the coastal dunes. Apart from mice, the land of the Falkland Islands supports precious little prey that sustained the warrah, but it is possible that insect larvae and pupae featured prominently in its diet. Although the interior of the Falkland Islands is rather impoverished when it comes to carnivore food, the coast is a bounteous source of nourishment at certain times of the year. The islands are used by numerous marine animals, including seals, sea lions, penguins, and a variety of flying seabirds. When these animals were raising their young, times must have been good for the warrah, and it probably made off with eggs, nestlings, adult birds, and even young pinnipeds. To reach these good supplies of food, the warrah traveled along well-worn paths that must have been made by generations of the animals accessing their feeding grounds via the shortest possible route. Although the southern spring and summer was a time of abundance for the warrah, the autumn and winter were probably very tough, and some accounts from the eighteenth and nineteenth centuries report that the living animals looked starved and very thin.

Regardless of its wintertime depravations, the warrah, in the absence of competition, appears to have been a successful species that was quite numerous on the two main islands of the Falklands group. This monopoly came to an end with the arrival of humans. Initially, visitors to the Falkland Islands were afraid of the warrah as it would wade into the water to meet an approaching boat. This was not an act of aggression, but an act of curiosity. The warrah had probably never seen humans and had therefore never learned to be afraid of them, an unfortunate fact that contributed to the extinction of this interesting dog.

Although the Falkland Islands are a harsh place, certain breeds of hardy sheep were well suited to the conditions, and they were introduced to the islands as a way of laying the foundations for the first human colonies on the islands. The sheep thrived on the islands, and as humanity tightened its grip on the Falklands, the warrah was seen as a menace that had to be exterminated. Like all dogs, the warrah was an opportunistic feeder, and it undoubtedly fed on the introduced sheep and lambs that nibbled the Falkland Island grass, but islanders, in their ignorance, believed the warrah was a vampire that killed sheep and lambs to suck their blood, only resorting to meat eating in times of desperation. Horrific myths can be very compelling, especially on a group of small islands where news travels fast and where livelihoods are at stake. In an attempt to quell the populace, the colonial government of the Falkland Islands ordered a bounty on the warrah, and fur hunters soon moved in to collect handsome rewards for delivering the pelts of dead animals.

The Falkland Islands, with a land area roughly the size of Connecticut, could never have supported huge numbers of warrah. Even before the human invasion, the warrah population was probably no more than a few thousand individuals, and it is therefore no surprise to learn that hunting quickly led to the extermination of this animal. Because the warrah was so very tame, hunting was a breeze, and all the hunter needed was a piece of meat and a knife. He held out the piece of meat to tempt the animal and stabbed it with the knife when it came within range. Other hunters used rifles or poison, but regardless of which particular method was used to kill the warrah, it was exceedingly rare by the 1860s.

Amazingly, a live warrah found itself in London Zoo in 1868 after being transported on a ship with a menagerie of other exotic animals, most of which perished during the journey. This warrah, far from home, survived for several years in the zoo, but it was one of the last of its species. Back in the South Atlantic, the onslaught of the sheep farmers and the hunters was too much for the poor warrah, and in 1876, the last known animal was killed at Shallow Bay in the Hill Cove Canyon.

+ The origins of the warrah are a mystery. Did it evolve on the Falkland Islands, surviving as a relic from the time before the last glaciation, when the islands were forested and home to a number of other land animals? Were the ancestors of the warrah brought to the islands by South American Indians as pets? Did the ancestors of the warrah walk to the Falkland Islands thousands of years ago when sea levels were much lower? Unfortunately, the answers to these questions died with the warrah, and the one-time presence of this canine in the South Atlantic remains a tantalizing zoological mystery.
+ Charles Darwin saw the warrah during his time on the *Beagle*, and it was clear to him that the species would not survive for very long in the face of human persecution. In actual fact, the warrah was exterminated in Darwin's own lifetime.
+ It was once a widely held myth that wolves sucked the blood of their prey, a belief that led to their persecution wherever they were found.

Further Reading: Alderton, D. *Foxes, Wolves, and Wild Dogs of the World*. Poole, UK: Blandford Press, 1994; Nowak, R. *Walker's Carnivores of the World*. Baltimore: Johns Hopkins University Press, 2005.

GREAT AUK

Scientific name: *Pinguinus impennis*
Scientific classification:
   Phylum: Chordata
   Class: Aves
   Order: Charadriiformes
   Family: Alcidae
When did it become extinct? The last pair of great auks was killed in 1844, although there was a later sighting of the bird in 1852 on the Grand Banks of Newfoundland.
Where did it live? The great auk was a bird of the Northern Atlantic, frequenting islands off the coast of Canada, Greenland, Iceland, and northern Europe.

In the roll call of recently extinct animals there is a long list of bird species, and flightless birds feature very prominently—hit hard by the spread of humans to the far reaches of the

Great Auk—The largest of the auks was killed off by overzealous hunting. (Natural History Museum at Tring)

globe. Often, these birds were giants of their kind, and the great auk, as its name suggests, was no exception. The Northern Hemisphere's version of the penguin, the great auk was a large bird that stood around 75 cm high and weighed about 5 kg when fully grown. Like the other auk species, the great auk had glossy black plumage on its back and head, while its underside was white. In front of each eye was a white patch of plumage.

Although the wings of the great auk were rather short and stubby, they were used to great effect underwater, where they would whirr away to propel the animal forward very rapidly through this dense medium. Like all auks and the unrelated penguins, the great auk was very maneuverable underwater, and it would pursue shoals of fish at high speed, seizing unlucky individuals in its beak. From remains of its food that have been found off the coast of Newfoundland, we know that the great auk hunted fish that were up to about 20 cm long, including such species as the Atlantic menhaden and the capelin. The grace and ease with which the great auk sliced through the water was not reflected in the way it moved about on land. It was built for swimming, and on land it was a very cumbersome animal, waddling around in the same way as the larger penguin species. As its feet were positioned far back on its body, it shuffled around and may have resorted to hops or sliding on its belly to overcome small obstacles. The ungainliness of the great auk on land was undoubtedly one of its downfalls because it could be caught with such ease.

Birds, no matter how well adapted they are to an aquatic existence, are always tied to the land. They need to return to land to lay their eggs and rear their young. During the breeding season, the great auks made use of low-lying islands to mate and lay their eggs. The female great auk only laid one egg per season, directly onto the bare rock. The egg was quite a specimen, weighing around 330 g. Every egg in the breeding colony was patterned slightly differently so that parents could easily recognize their own developing youngster. The parents probably fed the hatchling on regurgitated fish collected during frequent fishing trips, and on this diet rich in proteins and fats, the young grew quickly. They had to, as the summer in these northern climes is very short indeed, and if the young hadn't grown sufficiently to take to the sea when the harsh conditions of winter descended, they would have perished.

Life for the great auk was tough, and it got a whole lot tougher when they caught the attention of humans. Europeans soon realized the great auk represented a treasure trove of oil, meat, and feathers. Their awkwardness on land coupled with an obligation to form dense breeding colonies on low-lying islands made them easy pickings for Atlantic mariners. Sailors armed with clubs would land on the breeding islands and run amok through the nesting birds, dispatching them with blows to the head. There are stories of great auks being herded up the gangplanks of waiting ships and being driven into crudely constructed stone pens to make the slaughter even easier. Once killed, the birds were sometimes doused in boiling water to ease the removal of their feathers. The plucked bodies were then skinned and processed for their oil and meat. The oil was stored and taken back to the cities of Europe, where it was used as lamp fuel, whereas the feathers and down from the bird were used to stuff pillows. The slaughter was relentless, and as breeding pairs of the great auk could only produce one egg per year, the species was doomed. It is known that the populations of great auk off the coast of Norway were extinct by 1300. By 1800, the last large stronghold of this bird, Funk Island, was targeted by hunters, and the great auk was effectively on a headlong course for extinction. The island of Geirfuglasker, off the coast of Iceland, was the last real refuge for this bird as it was inaccessible; however, the island was inundated with water during a volcanic eruption and an earthquake. The birds that survived fled to the island of Eldey, near the tip of the Reykjanes Peninsula, Iceland, and it was here that the last breeding pair was killed on July 3, 1844, by two Icelanders. This last pair of great auks was killed while brooding an egg, and this, the last egg laid by the great auk, was smashed. Lonely individuals of the great auk may have scoured the North Atlantic looking for others of their kind as one was apparently spotted around the Grand Banks in 1852, but their searches were in vain, and they, too, eventually went the same way as the rest of their species.

- The great auk was just one species of a number of giant, flightless auks that inhabited the Atlantic. All of them, except the great auk, became extinct several thousand years ago.
- The great auk's similarity in both appearance and lifestyle to the penguins of the Southern Hemisphere is a very good example of convergent evolution, the phenomenon whereby two unrelated species come to resemble each other as a result of having to adapt to similar environments.

+ Bones from archaeological sites in Florida suggest that the great auk may have migrated south over the winter to escape the worst of the weather.
+ The museums of the world hold many great auk remains. There are numerous skins—many of which have been used to create stuffed reconstructions—eggs, and bones. However, complete skeletons of the great auk are very rare, with only a few known to exist. The eyes and the internal organs from the two last known great auks were removed and preserved in formaldehyde. These poignant reminders of the extinction of this fascinating animal can be seen in the Zoological Museum in Copenhagen, Denmark.

Further Reading: Olson, S. L., C. C. Swift, and C. Mokhiber. "An Attempt to Determine the Prey of the Great Auk (*Pinguinus impennis*)." *Auk* 96 (1979): 790–92; Fuller, E. *The Great Auk*. New York: Abrams, 1999.

Extinction Insight: The Great American Interchange

Following their discovery by Europeans in 1492, North America and South America have been collectively known as the Americas or the New World, two immense landmasses that had been close geographical neighbors for time immemorial. However, the geological histories of North and South America are very different, and for huge expanses of time, there has been no physical link between them whatsoever. All of the landmasses on earth were once assembled in a superlandmass, Pangea. Over millions of years, Pangea fragmented, and all of the continents in the modern Southern Hemisphere were grouped as a southern supercontinent, Gondwanaland, while the continents of the Northern Hemisphere formed the northern supercontinent, Laurasia. Over millions of years, these supercontinents were wrenched apart by the colossal forces of plate tectonics into the landmasses we are familiar with today, and they were rafted over the viscous rock of the earth's mantle to more or less their current positions. Although South America faced North America across the equator, there was no physical connection between the two landmasses.

Great American Interchange—The emergence of a land bridge between North and South America allowed animals to migrate between these two landmasses. Several types of North American mammal moved into South America, but relatively few of the South American mammals made it to the north and thrived. (Phil Miller)

North America retained a connection to the other landmasses by way of the intermittent land bridges that formed between its northwestern corner and the eastern tip of Asia. South America, on the other hand, has been completely isolated during its history for immense stretches of time.

The animal inhabitants of South America evolved in isolation to form a fauna that was amazing and unique. The mammals were particularly interesting, and many groups were known only from South America. Although South America was isolated from the other landmasses, some animals managed to set up home there by inadvertently rafting across the then narrow Atlantic Ocean from Africa on floating mats of vegetation. This is how rodents and monkeys are thought to have reached South America between 25 and 31 million years ago. Much later, at around 7 million years ago, some representatives of the group of mammals that includes raccoons and coatis managed to reach South America from North America using stepping stones of islands that were appearing between the two landmasses. These islands were the highest reaches of modern-day Central America, which was being uplifted from below the waves.

The isolation of South America and the uniqueness of its fauna was upset completely about 3 million years ago when the gradual geological upheaval forced the Isthmus of Panama out of the ocean completely, directly connecting the two landmasses. This was the beginning of the Great American Interchange and over the next few thousand years, animals and plants used the corridor of dry land to move between North America and South America. Many species of mammal we associate with South America actually originated in North America, for example, the llamas and tapirs. Other migrants from the north included horses; cats such as the cougar and jaguar; dogs; bears; and several types of rodent, to name but a few animals. Some South American mammals managed to cross the land bridge into North America, but many of these are now extinct, including the glyptodonts and giant ground sloths. The only surviving North American mammals to have their origins in South America are the Virginia opossum, the nine-banded armadillo, and the North American porcupine.

For reasons that are not completely understood, the South American species did not fare well when it came to invading the north, while the North American species thrived in the South American lands. The only ancient South American animals to make any lasting impression in North America were the ones with some sort of protection. The extinct glyptodonts, like the armadillos, were protected with a tough carapace, while the ground sloths had powerful claws, thick skin, and great size on their side. Apart from mammals, one other group of South American animals, the terror birds, managed to survive in North America for a while, but it is possible that they crossed by island hopping before the two landmasses became connected by a corridor of land.

The animals that moved into South America from the north thrived, and most of them are still around today, even though this continent has been massively altered by humans. All of the South American cats, bears, and dogs have their origins in North America, but they all adapted to the varied habitats offered by this continent and may have even played a role in driving some of the South American native mammals to extinction. The giant, native animals that were unique to this continent are all extinct, and all that we have as reminders of their existence are dry bones and a few pieces of parched hide. Although the original South American giants are all gone, their smaller relatives live on. Today, more than 80 species of marsupial survive in South America, but they are mostly tree-dwelling animals with a liking for insects and fruit. The relatives of the giant ground sloths live on in the trees as the five species of forest sloth, famous for their sluggish behavior. The anteaters, strikingly different to all other mammals, are not unique to South America, but it is here they reach their greatest size in the shape of the giant anteater. Superficially similar to the glyptodonts, the armadillos live on as 20 living species, but they are distantly related to the armored giants of the Pleistocene, which grew to the size of a small car.

Many hundreds of thousands of years after the Great American Interchange reached its peak, humans moved into the Americas via the Bering land bridge, although there is increasing evidence that early seafarers may have reached these lands a long time before people walked across. Regardless of how humans got to North America, they also moved south into South America. Early crossings may have been made using boats, but the land bridge used by the animals of the New World for millennia was certainly used by humans as well.

3

FEWER THAN 500 YEARS AGO

ELEPHANT BIRD

Elephant Bird—The largest of the elephant bird species weighed around 450 kg. (Renata Cunha)

Scientific name: *Aepyornis* sp.
Scientific classification:
    Phylum: Chordata
    Class: Aves
    Order: Struthioniformes
    Family: Aepyornithidae
When did it become extinct? It is not precisely known when the elephant bird became extinct, but it may have hung on until the eighteenth or nineteenth century.
Where did it live? The elephant bird was found only on the island of Madagascar.

Elephant birds were among the heaviest birds that have ever existed. Following the extinction of the last dinosaurs 65 million years ago, the mighty reptiles that had dominated the earth for more than 160 million years, the long over-shadowed birds and mammals evolved into a great variety of new species, some of which gave rise to giants like the elephant bird.

In their general appearance, elephant birds were similar to the flightless birds called "ratites" with which we are familiar today, such as the emu (*Dromaius novaehollandiae*), ostrich (*Struthio camelus*), rhea (*Rhea* sp.), cassowary (*Casuarius* sp.), and kiwi (*Apteryx* sp.); however, the biggest elephant bird, *Aepyornis maxiumus*, was enormous. It was about 3 m tall and probably weighed about 450 kg (the giant moa of New Zealand was actually taller but was way behind the elephant bird in terms of bulk—moa are discussed later in this chapter). On the island of Madagascar, there were few large predators, and the ancestors of the elephant birds had no need to fly; therefore this ability was gradually lost. Grounded, these birds went on to become animals that were bound to the land. Their skeletons show that they had very powerful legs and that they plodded around Madagascar on their big feet. The wings were reduced to tiny structures and were probably not visible beneath the bird's plumage. These birds had become so well adapted to a life without flight that the large and specially modified chest bone (keellike sternum) found in most birds, which serves as an attachment for the wing muscles, had all but disappeared.

We don't know exactly what the elephant birds ate, but we can assume from the shape of their bill that they were not carnivorous. Some people have suggested that certain Madagascan plants that are very rare today depended on the elephant birds for the dispersal of their seeds. The digestive system of these large birds was ideally suited to breaking down the tough outer skins of these seeds. Some were digested, but others passed through the bird intact and in a state of readiness for germination.

The remains of the elephant bird that have been found to date allow us to build up a picture of how this extinct animal lived. The most intriguing remains are the bird's eggs. Some have been found intact, and they are gigantic—the largest single cells that have ever existed. They are about three times bigger than the largest dinosaur eggs, with a circumference of about 1 m and a length of more than 30 cm. One of these eggs contained about the same amount of yolk and white as 200 chicken's eggs. These huge shelled reminders of the elephant bird are occasionally unearthed in the fields of Madagascan farmers, and one is even known to contain a fossilized embryo.

The number of elephant bird species that once inhabited Madagascar is a bone of contention among experts, but it is possible that Madagascar supported several species of these large birds. On their island, surrounded by abundant food and few animals to fear, especially when fully grown, the elephant birds were a successful group of animals. Then, around 2000 years ago, their easy existence was overturned as humans from Africa, Indonesia, and the islands around Australia reached this isolated land of unique natural treasures. Humans by themselves are one thing, but thousands of years ago, humans did not travel alone—they took their domestic animals with them. The elephant birds, in their 60 million years of evolution, never saw a human, and they wouldn't have recognized them as dangerous. The humans, on the other hand, saw the elephant birds as a bounteous supply of food. Hunting had a disastrous effect on the populations of these giant birds. They had evolved in the absence of predation and, as a result, probably reproduced very slowly. To add insult to injury, the animals the humans brought with them—pigs, dogs, rats, and so on—made short work of the elephant bird's eggs. Other introduced animals, such as chickens, may have harbored diseases to which these giant birds had never been exposed. With no natural immunity to these pathogens, epidemics may have ravaged the populations of elephant birds, which were already under pressure from hunting and egg predation. Changes in climate may have led to the drying out of Madagascar, and this, too, could have affected the populations of these impressive birds. The actual extinction timeline for the elephant birds is sketchy, but many experts suppose that the last of these great birds died out before 1600. The means at our disposal for the aging of ancient material are constantly improving, and some recent estimates move the disappearance of these birds into the nineteenth century. It is possible that some stragglers managed to survive until recent times, but we can be certain that no elephant birds survive today.

- The Island of Madagascar was once part of Africa, but over millions of years, the tectonic forces of continental drift rafted it away from the African mainland and into the Indian Ocean. The animal inhabitants of this huge island evolved in isolation to produce animal and plant species that were very different from those found elsewhere. Although the elephant birds are all extinct, Madagascar is still home to many other unique animals—the most notable of these being the lemurs.
- The elephant bird has always been shrouded in myth and legend. In the thirteenth century, the great explorer Marco Polo recounted tales of a huge bird of prey that could carry an elephant in its huge talons. Known as the roc or rukh, the stories of this bird convinced sailors who visited Madagascar and saw eggs of the elephant birds that the island was home to this giant raptor. This is where the name "elephant bird" may have come from, and it appears to have stuck, even when Europeans realized that the elephant bird was actually like a giant ostrich.
- Memories of the elephant bird persisted for a long time in the stories and histories of some of the native Madagascan people (Malagasy). These stories describe the elephant birds as gentle giants. Although these accounts are liable to exaggeration, it gives us some idea of what the living elephant bird may have been like.

Further Reading: Cooper, A., C. Lalueza-Fox, S. Anderson, A. Rambaut, and J. Austin. "Complete Mitochondrial Genome Sequences of Two Extinct Moas Clarify Ratite Evolution." *Nature* 409 (2001): 704–7; Goodman, S.M., and J.P. Benstead, eds. *The Natural History of Madagascar.* Chicago: University of Chicago Press, 2003.

STELLER'S SEA COW

Scientific name: *Hydrodamalis gigas*
Scientific classification:
  Phylum: Chordata
  Class: Mammalia
  Order: Sirenia
  Family: Dugongidae
When did it become extinct? It became extinct in the year 1768, although it is possible that the species may have persisted for a few more years.
Where did it live? The last populations of Steller's sea cow were known from some of the islands in the Bering Sea, just off the coast of the Kamchatka Peninsula.

In 1741, the *St. Peter,* captained by Vitus Bering, departed from Kamchatka. The mission was to find an eastern passage to North America. On board was a 32-year-old German

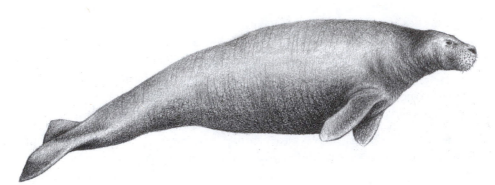

Steller's Sea Cow—At least 8 m long, Steller's sea cow was the largest marine animal apart from the whales, and it is the largest animal to have gone extinct in relatively recent times. (Phil Miller)

by the name of Georg Wilhelm Steller, who was the ship's official mineralogist. Steller also happened to be a physician and a very keen naturalist. His journey on the ship through the Bering Sea would be a remarkable one, on which he would make many zoological discoveries. Steller diligently documented everything he saw, and most of what we know about Steller's sea cow is thanks to the notes he and a crew mate, Sven Waxell, made in their journals.

The sea cows were observed around Bering Island and Copper Island, where they could be observed floating among and feeding on the vast marine forests of kelp that grew in the shallows around these islands. Steller's observations give us an insight into how this animal lived and what it looked like. Steller's sea cow was a huge animal and one of the biggest creatures to have become extinct in very recent times. It was very closely related to the dugongs and manatees, the unusual marine animals found in tropical rivers, estuaries, and shallow marine habitats around the world, but it was very much larger. Adults could grow to around 8 m, and the great bulk of the animal suggests weight in excess of 4,000 kg—possibly over 8,000 kg. They were gentle animals that apparently spent their time grazing on kelp—leaving great mounds of the seaweed washed up on the shore—and snoozing. In place of teeth, they had a bony ridge in their upper and lower jaws to grind the fibrous algae, and their forelimbs were stout flippers, which the animals could use to provide purchase on the rocky seabed when they were feeding in the very shallow coastal water. The animals' skin was rugged, thick, and black, and Steller likened it to the bark of an old tree. The downfall of Steller's sea cow was its flesh—a valuable commodity to the crew of the *St. Peter*, who were shipwrecked on Bering Island. Not only were these huge marine animals slow moving and gentle, but they also lived in family groups and appear to have been very curious. Steller observed them investigating the small boats of men who carried guns and spears to shoot and stab them. In what was a very wasteful strategy, the wounded animals were allowed to swim off in the hope that the surf and tide would bring them ashore. Often this was not the case, and the moribund animal would simply die and sink. The animals that were landed were butchered, and although the flesh had to be boiled for quite some time, it was very similar to beef in taste. When the survivors of the *St. Peter* were rescued along with barrels of Steller's sea cow meat, it was not long before whalers, fishermen, and hunters, attracted to the area for the bounteous amount of wildlife, turned their attention to these gentle animals to nourish them on their expeditions. Not only did they eat the meat and fat of this animal, but the oil from its blubber was also coveted because it gave off little smoke and odor when it was burned. The skin was processed to make a range of leather goods.

It has been suggested that even when Steller first observed the sea cow in 1741, it was already rare, its populations reduced to a fraction of their former strength by human hunting over thousands of years. Indeed, bones and fossils show that this species lived along much of the North Pacific coast, from Baja northward and down to northern Japan. What Steller discovered were the last populations of this impressive animal, which had survived in a remote, inhospitable area. As it was such a large animal, it is very likely that Steller's sea cow was a slow breeder, a fact that made it even more vulnerable to the effects of overhunting. Whatever the state of the population of this animal when it was discovered, we know that by 1768, 27 years after it was described by Steller, it was extinct. It is possible that a few individuals survived in the shallow waters of other islands in the Bering Sea, but an expedition in the late eighteenth century did not find any sea cows. Even today, some people cling to the hope that Steller's sea cow survived into the modern day, with claims of sightings around the islands in the Bering Sea. Unfortunately, it is highly unlikely that such a large animal, which spent so much of its time at the surface, has escaped detection in an increasingly crowded world. Twenty-seventy years is an amazingly short amount of time for an animal to be wiped out, and it shows just how relentless humans can be in their extermination of other creatures.

- Steller, during his time on the *St. Peter*, documented hundreds of new species, including the northern fur seal (*Callorhinus ursinus*), the sea otter (*Enhydra lutris*), Steller's sea lion (*Eumetopias jubatus*), Steller's eider duck (*Polysticta stelleri*), and the spectacled cormorant (*Phalacrocorax perspicillatus*). All except the last species can still be seen today, but the populations of them all suffered terribly at the hands of hunters, who streamed into the area after Bering's ill-fated voyage. The spectacled cormorant, a large marine bird with a distinct unwillingness to take to the wing, was last seen around 1850.
- Along with the species that now bear his name, Steller also recorded other animals that have never been verified. One of these was described by him as the "sea ape," a marine animal with an unusual collection of features. It is impossible to know if the sea ape and others are animals we know today, but Steller's documented observational abilities leave us with the tantalizing possibility of other, as yet unknown animals swimming in the cold but productive waters of the Bering Sea.
- In December 1741, the *St. Peter* was forced to seek refuge from the atrocious conditions in the Bering Sea on what became known as Bering Island. Vitus Bering died of scurvy on this island, along with 28 of his crew. The survivors, with Steller among them, saw out the winter; they constructed a new vessel from the remains of the *St. Peter* and returned to Kamchatka. Back on the mainland, Steller spent the next two years exploring the vast peninsula of Kamchatka, documenting its animals, plants, and geology. He was eventually requested to return to St. Petersburg but died of an unknown fever on his way back.

Further Reading: Anderson, P. "Competition, Predation, and the Evolution and Extinction of Steller's Sea Cow *Hydrodamalis gigas*." *Marine Mammal Science* 11 (1995): 391–94; Scheffer, V.B. "The Weight of the Steller Sea Cow." *Journal of Mammalogy* 53 (1972): 912–14.

DODO

Dodo—Although the dodo is one of the most well known recently extinct animals, very few remains of this animal survive to this day. (Renata Cunha)

Scientific name: *Raphus cucullatus*
Scientific classification:
  Phylum: Chordata
  Class: Aves
  Order: Columbiformes
  Family: Columbidae
When did it become extinct? The dodo is generally considered to have gone extinct in 1681, but any records of it after the 1660s have to be treated with caution.
Where did it live? The dodo was only found on the island of Mauritius, 900 km to the east of Madagascar.

"As dead as a dodo!" No phrase is more synonymous with extinction than this one. The dodo is the animal that springs to mind when we think of extinction. Often portrayed as a stupid, bumbling giant of a bird, the dodo was actually a very interesting animal that was perfectly adapted to its island habitat. Unfortunately, its evolutionary path had never counted on humans; thus, when we discovered these birds, they didn't last very long.

We don't know exactly what the dodo looked like as no complete skin specimen exists, but we do know it was a large bird, about the same size as a large turkey, with a stout build, sturdy legs, thick neck, and large head. Fully grown specimens were probably around 25 kg in weight and as tall as 1 m. The dodo's most characteristic feature was its very large beak (up to 23 cm long), complete with bulbous, hooked tip. The wings were stubby and effectively useless as the dodo evolved on an island where there were no predators, and therefore flight was an expensive waste of energy; instead, it ambled about on the forest floor of its Mauritian home. The only information we have on what the dodo ate is from the accounts of seafaring people who stopped off on the island of Mauritius and saw the bird going about its everyday business. The favored food of the dodo was probably the seeds of the various Mauritian forest trees, but when its normal source of food became scarce in the dry season, it may have resorted to eating anything it could find. A liking for seeds ties in with other observations of the dodo's behavior, which report that it ate stones. These stones passed into the dodo's crop, which is like a big, muscular bag, and there they assisted in grinding the hard-shelled seeds.

As the dodo couldn't fly, it could only build its nest on the ground. Sailors described these nests as being a bed of grass, onto which a single egg was laid. The female incubated the egg herself and tended the youngster when it hatched. Sailors who saw the living birds said the young dodo made a call like a young goose. Apart from small pieces of information, we know very little about the behavior of the dodo. We have no idea if they lived in social groups or how the adults interacted during the breeding season. What we do know is that they were hopelessly ill adapted to deal with human disturbance.

The dodo was first described in 1598, although Arab voyagers and Europeans had discovered Mauritius many years previously and had undoubtedly seen its unique animals. The large dodo excited hungry seafarers who had not eaten fresh meat for many months while out at sea; however, the flesh of the dodo was far from flavorsome. Even the unpleasant taste of the dodo's tough flesh didn't stop people from killing them for food, often in large numbers, and any birds that could not be eaten straight away were salted and stored on the ship for the rest of the voyage. Hunting the dodo was said to be a very easy exercise. It couldn't fly or even run at any great speed, and it also had the great misfortune of being completely unafraid of humans. Dodos had never seen a human, and as a result, they had not learned to be afraid. It is said they would waddle up to a club-wielding sailor only to be dispatched with one quick swipe. In the rare situation in which they felt threatened, they would use their powerful beak to good effect and deliver a painful nip.

Hunting obviously hit the dodos hard—their size and small clutches suggests that they were long-lived, slow-breeding birds, which was not a problem in the absence of predators, but as soon as humans and their associated animals entered the equation, extinction was inevitable. Seafarers who visited Mauritius brought with them a menagerie of animals, including dogs, pigs, rats, cats, and even monkeys. These animals disturbed the nesting dodos and ate the lonesome eggs. With this combination of hunting, nest disturbance, and egg predation, the dodo was doomed. It has been suggested that flash flooding could have tipped the dodo population, already ravaged by hunting, nest disturbance, and egg predation, over the edge into extinction. Regardless of the causes, the enigmatic dodo was wiped out in a little over 100 years after it was first discovered by Europeans.

+ The dodo is in the same group of birds that includes the doves and pigeons. Its ancestor was probably a pigeonlike bird that alighted on the island of Mauritius, evolving over time into a big, flightless species.

+ The last record of the dodo is commonly said to be that of an English sailor, Benjamin Harry, who visited the island in 1681. This and other late records of the dodo are thought to refer to another extinct Mauritian bird—a type of flightless rail called the "red hen." Historically, it was common for the name of an extinct animal to be transferred to another species living in the same location.

+ Rodrigues Island, 560 km to the east of Mauritius, was once home to another species of big, flightless bird. This bird, known as the Rodrigues solitaire (*Pezophaps solitaria*), was first recorded in 1691, yet by the 1760s, at the very latest, it, too, had gone the same way as its relative, the dodo. Réunion Island, also in the Mauritius group, was said to be the home of a completely white dodo called the "Réunion solitaire"; however, it has now been established that this bird was actually an ibis, rather than a dodo. Sadly, this bird is also extinct. Albino dodos were actually observed on Mauritius and undoubtedly added to the confusion over the identity of the Réunion solitaire.

+ Although the dodo is a very familiar extinct animal, remarkably few remains of it exist in collections. There are a few complete skeletons, a few disjointed bones, and a head and foot that still have tissue attached. The foot and head came from the last stuffed specimen, which was once on display in the Oxford Ashmolean Museum. Apparently, by 1755, the specimen was in quite a sorry state, and it was said that the curator ordered it to be burned. This recklessness is now thought to be a myth and the burning was, in fact, a desperate attempt by museum workers to salvage what they could from a badly disintegrating specimen, leaving us with the remnants we have today.

+ Mauritius and its neighboring islands, thanks to their isolation in the Indian Ocean, were home to many species of unique animal before the arrival of Europeans and the destructive animals they had in tow. At least we have a good idea of what the dodo and Rodrigues solitaire looked like—unfortunately, the same cannot be said for many of the other animals with which these birds shared their home. We now know these islands harbored other flightless and flying birds, bats, giant tortoises, and even snakes, all of which are now extinct. Precious little information is available on these animals.

Further Reading: Cheke, A. S. "Establishing Extinction Dates—The Curious Case of the Dodo *Raphus cucullatus* and the Red Hen *Aphanapteryx bonasia*." *Ibis* 148 (2006): 155–58; Johnson, K. P., and D. H. Clayton. "Nuclear and Mitochondrial Genes Contain Similar Phylogenetic Signal for Pigeons and Doves (Aves: Columbiformes)." *Molecular Phylogenetics and Evolution* 14 (2000): 141–51.

AUROCHS

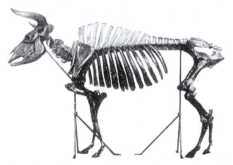

Aurochs—The aurochs was the ancestor of most modern cattle, albeit significantly larger than most modern breeds. Both males and females feature prominently in ancient cave art. (Cis Van Vuure)

Scientific name: *Bos primigenius*
Scientific classification:
  Phylum: Chordata
  Class: Mammalia
  Order: Artiodactyla
  Family: Bovidae
When did it become extinct? The last known aurochs died in 1627.
Where did it live? The aurochs was found throughout Europe, the Middle East, and into Asia, with subspecies in North Africa and India.

Most of the cattle breeds we know today are descended from the huge prehistoric cattle known as aurochs. These large animals roamed the woods and glades of Europe and Asia for thousands of years, until the last of the species, a female, died in Poland in 1627.

As the aurochs only disappeared in quite recent times, there are lots of accounts of what it looked like and how it behaved. The males were very large animals—1.8 m at the shoulder and 900 kg—significantly larger than most of the cattle breeds we have today. Both the males and females had impressive horns that curved forward and slightly inward, and the male in particular looked like a typical but very powerfully built bull. Unlike modern breeds of cattle, the male and female aurochs were a different color. A bull was said to be black with a pale stripe along his spine, while the female was more reddish brown.

Aurochs—This old drawing, by an unknown artist, clearly shows the distinctive horns of the aurochs. (Cis Van Vuure)

According to historic accounts, aurochs lived in family groups that were made up of females, calves, and young bulls. As the bulls grew older, they formed groups of their own, and the large, mature bulls were solitary, only mixing with others of their kind during the breeding season. Like other types of cattle, the aurochs were completely herbivorous and lived on a diet of grasses, leaves, fruits such as acorns, and even the bark of trees and bushes during the harsh winter months.

The aurochs, particularly the bulls, were said to be very aggressive, and they were apparently very difficult to domesticate, but about 9,000 years ago, in the Middle East, early humans did exactly that, giving us many of the cattle breeds we have today. A large animal with an aggressive nature would not have been easy to look after, so our ancestors selectively bred these animals to make them more docile. Selective breeding was also used to produce types of cattle that could yield copious amounts of milk. The udders of the female aurochs were far smaller than the capacious glands in between a modern cow's back legs.

Humans domesticated many other animals apart from the aurochs, and it was this change from a hunter-gatherer existence to an agricultural one that spelled the end for the aurochs. Over centuries and millennia, humans changed the habitats in which the aurochs lived. They cut down the forests to plant crops or to make room for their domesticated animals to graze and browse. The land they chose for their first agricultural attempts were those areas with the richest soils: river deltas, valleys, and fertile wooded plains. These were the aurochs' natural habitat, and they were forced into areas where the food was perhaps not quite as nutritious. The large size and formidable temperament of these animals made them very popular hunting targets for food and sport. Habitat loss, competition with their domesticated relatives, and hunting all contributed to the gradual disappearance of the aurochs. In 1476, the last known aurochs lived in the Wiskitki and Jaktorów forests, both of which are in present-day Poland. These last two populations of aurochs were owned by the Duke of Mazovia, and as they were favored animals for hunting, they eventually received royal protection, making it an offense for anyone other than a member of the royal household to kill an aurochs. Unfortunately, what is now Poland fell into turbulent times, and many kings came and went in quite a short period of time. During this era, the protection of the aurochs was much less of a priority, and the last two populations got smaller and smaller through neglect and hunting. From 1602, records show that aurochs were only found in Jaktorów Forest, and a royal decree was issued in 1604 to protect the remaining individuals. This was too little too late, and by 1627, the species was extinct—the forests of central Europe would no longer hear the bellow of an aurochs bull.

+ The ancestors of the aurochs are believed to have evolved in India around 1.5 to 2 million years ago, from which time they spread throughout the Middle East, Asia, and Europe. For much of their existence, the earth was going through ice ages and intervening warm periods, and as the aurochs were not adapted to survive in intensely cold environments, their range probably increased as the ice sheets withdrew and contracted as the ice sheets extended south.

+ The aurochs died before photography was invented, so we have no photographs, and considering that this was once a very common animal, there are not many complete skeletons in the world's museums. The image of the aurochs lives on in cave paintings, and the La Mairie cave (Dordogne, France) pictures, which date back to around 15,000 years ago, show a bull aurochs with two females.

+ In the 1920s, two German zoologist brothers speculated that the aurochs could be effectively brought back from the dead by selectively breeding modern cattle for aurochs traits. Their experiments quickly produced cattle with some strong similarities to the aurochs. These animals, known as Heck cattle, do have some of the characteristics of the aurochs, but they can only ever be an approximation of the extinct animal and an interesting experiment in selective breeding.

+ Some animal breeders and zoologists have suggested that the fighting bulls of Spain have many aurochslike characteristics and so perhaps they represent the closest living relatives of these extinct beasts.

+ There is an ongoing, intense debate on how Europe looked after the end of the last ice age. One group of scientists believes that all of Europe was covered by dense forest until humans came along and started chopping it all down. Another group supports the idea that feeding and trampling by large animals like the aurochs opened up and maintained large glades and paths within the forest. Białowieża Forest, a World Heritage Site and biosphere reserve on the border between Poland and Belarus, is the last remnant of this European wildwood.

Further Reading: van Vuure, T. "History, Morphology and Ecology of the Aurochs (*Bos primigenius*)." *Lutra* 45 (2002): 1–16.

MOA

Scientific name: Several species
Scientific classification:
    Phylum: Chordata
    Class: Aves
    Order: Struthioniformes
    Family: Dinornithidae

When did it become extinct? Estimates for the disappearance of the moa vary, but it is thought they became extinct in the 1500s.

Where did it live? The moa were found only in New Zealand.

The elephant birds (see the earlier entry in this chapter) were not the only giant birds that roamed the earth in quite recent times. Thousands of miles to the east of Madagascar, the islands of New Zealand were once home to several species of large bird, the largest of which was taller than the elephant birds, although much more slender. Collectively, these birds were known as moa (a Polynesian word meaning "fowl"), and they had lived on the islands of New Zealand for tens of millions of years. The only mammals that had managed to reach New Zealand were bats, so the islands were free of any large ground-based predators or herbivores, absences which allowed the ancestors of the moa to evolve in unique ways. First, as there were no mammalian predators, flight was an unnecessary extravagance, especially as food was so abundant. Flight limits the maximum size a bird can ever be, and so without this limitation, the moa grew to huge sizes. Second, as there were no mammalian herbivores in New Zealand, the moa evolved to fill this gap, taking on the ecological role that animals such as deer fill in many other parts of the world.

Moa—Several species of moa once inhabited the islands of New Zealand. They ranged in size from 1-m-tall, 25-kg birds to 4-m-tall, 275-kg giants. (Renata Cunha)

Today, between 10 and 15 species of moa are recognized by scientists from their remains, but it is impossible to know exactly how many species of these interesting birds once inhabited the islands of New Zealand. Some experts have suggested that there could have been as many as 24 species of moa. The smallest species were around 1 m tall and weighed around 25 kg, while the biggest species, *Dinornis robustus*, on South Island, and *Dinornis novaezelandiae*, on North Island, were enormous, at around 4 m at their full height and 275 kg in weight. Interestingly, moa skeletons and reconstructions are almost always shown standing upright, but scientists now think that they walked around with their neck held more or less horizontal to the ground, but they could have probably risen to their full height when they needed to. All the moa were covered in very fine feathers, resembling hair—much like the kiwis—and all of them had very robust legs ending in powerful, clawed feet. Much of the head, throat, and lower legs were featherless. The wings of the moa had become so useless that they had shrunk away to almost nothing and only remained as small vestigial flaps beneath the hairlike plumage.

The moa were all herbivores, and as they diversified into a range of species, they probably fed on different plants in different habitats. Some of the species may have grazed the plants in the lowlands, while other species nibbled low-growing herbs in the uplands. Although New Zealand was once free of mammalian predators, the moa did have an enemy in the shape of Haast's eagle (see the later entry in this chapter), an impressive aerial predator that probably assaulted the moa from the air and killed them with its powerful crushing talons. The only real defense the moa had against this predator were their powerful legs, which bestowed them with a good turn of speed when the need arose.

The bones and bits of mummified moa tissue that have been found tell us where the animal lived and what the animal looked like, but it can only partially illuminate the life of these long-dead animals. Like other birds, the moa laid eggs—big eggs (the biggest moa egg has the same capacity as about 100 chicken eggs)—and as building a nest up a tree was completely out of the question, these must have been deposited on the ground, probably in a simple scrape or on a mound of gathered vegetation. Unusually, the female moa was much larger that the male, and this suggests that they must have had some interesting breeding system the likes of which we can only guess, but it is reasonable to assume that the female protected a territory and attracted her suitors—a reversal of what is seen in many bird species, where the male has to attract mates.

What happened to these feathered giants? The simple answer is humans. Polynesians (called Māori), on their seafaring craft, reached New Zealand around A.D. 1300, and their effect on the plants and animals of these islands was dramatic. We can only imagine what these people thought when they reached New Zealand, but they must have been at sea for a long time without charts and no idea of their destination, so for them to come across these verdant, volcanic islands stocked with all sorts of food must have been cause for celebration. There is evidence to suggest that these migrants started wildfires, maybe to clear areas for the cultivation of crops or perhaps as a way of driving prey animals out from cover. They also hunted the moa directly, and what with the combined effects of this and habitat loss, the moa were doomed. The moa were probably long-lived birds, and it has been shown that they only reached full size at about the age of 10, with several more years passing before they reached sexual maturity. Therefore any factors that had an effect on the number of adults in the population, such as hunting and habitat loss, had a drastic effect on the population as a whole. It has recently been speculated that moa populations were on the decline before the arrival of humans, possibly due to disease transported by migrating birds gone astray or even due to explosive volcanic activity. Regardless of the possibility of a dwindling population, the moas were wiped out around 160 years following the arrival of humans—a startlingly short period of time and yet another demonstration of how destructive our species can be.

- It was once thought that the closet living relatives of the moa are the kiwis, but the current view is that they were more closely related to the emu of Australia and the cassowary of Australia and New Guinea.

- The ancestors of the moa are thought to have walked across to New Zealand when it was still part of the massive landmass known as Gondwanaland. Over tens of millions of years, tectonic forces rafted the lands of New Zealand apart until they became an isolated group of islands. The ancestors of the kiwis are thought to have flown to New Zealand after it had become separated.

- New Zealand is an oceanic archipelago that consists of two large islands, North Island and South Island, as well as many smaller islands. The land area and the diversity of the habitats on these islands provided the original inhabitants with a wealth of niches into which to evolve, and birds became the rulers of this realm.

- Since the arrival of humans to New Zealand, more than 58 species of native birds have become extinct.

- All birds evolved from small dinosaurs about 155 million years ago in the late Jurassic period. Ratites, the group of birds to which the moa belonged, evolved in Gondwanaland in what we know as South America. As this supercontinent was wrenched apart over millions of years into the landmasses with which we are familiar today, the ratites evolved into the moa and kiwis of New Zealand, the elephant birds of Madagascar (see the entry earlier in this chapter), the emu of Australia, the cassowary of Australia and New Guinea, the ostrich of southern Africa, and the rheas of South America.

Further Reading: Worthy, T. H., and R. N. Holdaway. *The Lost World of the Moa*. Bloomington: Indiana University Press, 2002; Turvey, S. T., O. R. Green, and R. N. Holdaway. "Cortical Growth Marks Reveal Extended Juvenile Development in New Zealand Moa." *Nature* 435 (2005): 940–43; Bunce, M., T. H. Worthy, T. Ford, W. Hoppitt, E. Willerslev, A. Drummond, and A. Cooper. "Extreme Reversed Sexual Size Dimorphism in the Extinct New Zealand Moa Dinornis." *Nature* 425 (2003): 172–75; Holdaway, R. N., and C. Jacomb. "Rapid Extinction of the Moas (Aves: Dinornithiformes): Model, Test, and Implications." *Science* 287 (2000): 2250–54.

HAAST'S EAGLE

Scientific name: *Harpagornis moorei*
Scientific classification:
    Phylum: Chordata
    Class: Aves
    Order: Falconiformes
    Family: Accipitridae
When did it become extinct? Haast's eagle is thought to have become extinct around 500 years ago, although it has been said that the species managed to survive into very recent times.
Where did it live? The eagle was found only in New Zealand.

Before the arrival of humans, birds ruled New Zealand. In the absence of mammalian predators, many of the feathered denizens of these islands gave up flying, and some of them evolved into giants such as the moa (see the entry earlier in this chapter). These islands were a treasure trove of animal prey for the animals that could reach them, and sometime between 700,000 and 1.8 million years ago, some small raptors, very similar to the extant little

Haast's Eagle—Haast's eagle was an enormous bird that was a specialist predator of New Zealand's extinct moa. (Renata Cunha)

eagle (*Aquila morphnoides*), were perhaps caught in a storm and blown off course, eventually finding themselves in the strange land of New Zealand, where their bird relatives quite literally ran the roost. This land was full of opportunity. Many of the native New Zealand birds were flightless herbivores and omnivores. There was a vacancy in New Zealand for an aerial predator that could tackle the numerous ground-dwelling birds, and the little lost eagle evolved rapidly to fill this niche. For much of the time, evolution moves at quite a slow pace, but if there's a space in an ecosystem, a species can evolve very rapidly to fill it. This is what happened with the ancestors of Haast's eagle, as a small bird of prey evolved into the largest eagle that has ever lived and the only eagle that has been the top predator in its ecosystem.

As with other top predators, Haast's eagle was probably never very common, and because of this, the remains of this fearsome predator are scarce. Three complete skeletons are known (the latest of which was discovered in 1990) as well as numerous fragmentary remains. The bones show just how big this eagle was. It has been estimated that a fully grown female weighed 10 to 15 kg and was 1.1 m tall, with a wingspan of around 2.6 m. This is approaching the limit of how heavy a bird dependent on flapping flight and maneuverability can be. For comparison, the heaviest living eagle, the harpy eagle (*Harpia harpyja*), weighs around 8 kg. The skull of Haast's eagle was around 15 cm long, but the bill was not as bulky as those of large, living eagles. Its claws are thought to have been tremendously powerful, and they were tipped with enormous, 7-cm-long talons.

For its size, Haast's eagle actually had short wings, a characteristic it shared with the harpy eagle. Many eagle species have long, broad wings, allowing them to soar effortlessly at high altitude for long periods of time, but in those species that have evolved in forest habitats, long wings would be a disadvantage. In these situations, stubbier wings are a much better bet, and because of this, it is thought that Haast's eagle was an animal of forests and bush.

With its great size, terrible talons, and maneuverability, Haast's eagle must have been a formidable predator, but what did it eat? At least a dozen moa skeletons have been found that bare gouges and scars in the bones of their pelvis. Until the arrival of humans, Haast's eagle was the top predator in New Zealand, and it is highly likely that the marks on these moa bones were caused during a predatory attack by Haast's eagle. From a perch in a tall tree, Haast's eagle surveyed its territory for moa and other large, ground-dwelling birds, and on sighting some suitable quarry, it launched an assault. Swooping toward the prey at a speed of between 80 and 100 km per hour, it swung its terrible talons forward in preparation for the contact. The eagle's powerful legs absorbed the force of the impact, but the prey was probably knocked clean off its feet. If the initial strike was not enough to kill the prey, the puncturing force of eight huge talons caused massive internal bleeding, and before long, the victim succumbed to blood loss and shock. With its prey dead, the eagle used its talons and beak to tear the skin of the hapless victim before digging into its flesh.

The large living eagles most often take prey that is considerably smaller than themselves so they can carry it away to a safe perch out of the way of scavengers. There were no scavengers in New Zealand large enough to challenge a Haast's eagle for its kill, and therefore it could tackle large prey and eat them where they died. At a kill, the only animals a Haast's eagle feared were others of its kind.

As formidable a predator as it was, the Haast's eagle was no match for humans, who first reached New Zealand around A.D. 1300. It is now a largely accepted theory that humans, through hunting and habitat destruction, brought about the extinction of the moa and many other unique New Zealand birds. Humans undoubtedly saw and knew this raptor, and whether they persecuted it or revered it is a bone of contention. In some cultures around the world, top predators are persecuted by humans, while in others, these animals are revered. Perhaps the Māori hunted Haast's eagle, not only because it competed with them for their food, but also as an act of reverence. In many aboriginal cultures, the body parts of powerful predators are collected and worn in the belief that the strengths of the animal will be transferred to the wearer. Hunting and dwindling prey probably killed off the Haast's eagle before the last moa disappeared.

- ✦ For a long time, it was assumed that Haast's eagle evolved from the wedge-tailed eagle that is found throughout Australasia. Recently, scientists managed to extract some DNA from Haast's eagle bones, and this was compared to the DNA of living eagles. This showed that the closest relative of Haast's eagle is the little eagle. Constructing a family tree from ancient DNA should always be done with caution as thousands of years lying in the ground can severely damage DNA, and old samples can be contaminated with DNA from sources too numerous to list.

+ A famous New Zealand explorer, Charles Douglas, a man who was not prone to exaggeration and flights of fancy, claimed in his journal that he had an encounter with two giant birds of prey in the Landsborough River Valley of South Island sometime in the 1870s. If this is true, is it possible that Haast's eagle somehow clung to existence in a remote part of New Zealand until very recent times? Unfortunately, we'll never know the truth as Douglas killed and ate both of these mysterious birds.
+ The bones of another giant raptor have also been found in New Zealand, and these are now thought to have once belonged to a massive type of harrier. Harriers are lightly built birds of prey weighing in at around 700 g. The New Zealand giant harrier (*Circus eylesi*) was more like 3 kg.

Further Reading: "Ancient DNA Tells Story of Giant Eagle Evolution." *PLoS Biology* 3 (2005): e20; Bunce, M., M. Szulkin, H.R.L. Lerner, I. Barnes, B. Shapiro, A. Cooper, and R.N. Holdaway. "Ancient DNA Provides New Insights into the Evolutionary History of New Zealand's Extinct Giant Eagle." *PLoS Biology* 3 (2005): e9; Brathwaite, D.H. "Notes on the Weight, Flying Ability, Habitat, and Prey of Haast's Eagle (*Harpagornis moorei*)." *Notornis* 39 (1992): 239–47.

MARCANO'S SOLENODON

Marcano's Solenodon—Marcano's solenodon, like its living close relatives, was a nocturnal predator of invertebrates and other small animals. (Phil Miller)

Scientific name: *Solenodon marcanoi*
Scientific classification:
    Phylum: Chordata
    Class: Mammalia
    Order: Soricomorpha
    Family: Solenodontidae
When did it become extinct? It is not known when Marcano's solenodon became extinct, but it was probably after the Europeans first reached the New World at the end of the fifteenth century.
Where did it live? The remains of this animal are only known from the island of Hispaniola.

Hispaniola, along with Cuba and Jamaica, make up the Caribbean island group known as the Greater Antilles. All these islands once had their own distinctive fauna, the ancestors of which somehow reached these islands from North, Central, and South America. Today, the native fauna of the Greater Antilles is a shadow of what it once was due to the arrival of humans: first, Amerindians, and much later, Europeans.

The solenodons have suffered badly at the hands of humans and their introduced animals. One species, Marcano's solenodon, a native of Hispaniola, is actually extinct and is only known from skeletal remains. Although we only have bones to work with, we can safely assume that Marcano's solenodon was very similar to the surviving solenodons in both appearance and lifestyle. Solenodons are a fascinating group of animals. In appearance, they look like large, well-built shrews and are about the same size as a very large brown rat (*Rattus norvegicus*), with reddish brown fur; a long, mobile snout; tiny eyes; and a long, scaly tail. The limbs of the solenodons are very well developed and the digits are tipped with long, sharp claws.

Like the living solenodons, Marcano's solenodon must have been a burrowing animal that only ventured from its daytime retreat to hunt and look for mates when night fell. This unusual animal was undoubtedly a carnivore, and the staple of its diet must have been insects and other invertebrates, including earthworms, centipedes, and millipedes, all of which were found by rooting through the leaf litter and the soil and by tearing up rotten logs on the forest floor. Marcano's solenodon was large enough to kill and eat vertebrates, such as small reptiles, amphibians, birds, and mammals, when the opportunity arose. One of the most fascinating things about the solenodon is its ability to secrete and use venom. Like almost all vertebrate venoms, solenodon venom is actually modified saliva. It is a mixture of various proteins produced by the salivary glands, and it is introduced into the body of the prey when the solenodon bites. The solenodons even have modified teeth for delivering this lethal cocktail. The large incisors on the lower jaw of the animal are equipped with a groove that channels the venom into the wound made by the teeth. Exactly how the solenodon venom kills the prey is unknown, but the venom produced by the American short tail shrew (*Blarina brevicauda*) causes the blood vessels to expand, leading to low blood pressure, paralysis, convulsions, and eventually, death. As the solenodons are closely related to the true shrews, we can assume that their venom has a similar effect.

Not only do the solenodons produce venom, but they also produce potent secretions from the base of their legs, which is said to have a strong, goatlike smell. Exactly what these secretions are for is unknown, but it is highly likely that they use this pungent aroma to mark their territory and communicate their willingness to mate to members of the opposite sex—such is their reliance on their sense of smell. Attracting mates with scent is important for a small, scarce mammal with poor eyesight. Exactly when these animals mate is unknown, but the females are receptive to the advances of males about every 10 days. When they meet, solenodons can be vocal animals, broadcasting their intentions with puffs, twitters, chirps, squeaks and clicks, but when the act of mating is over, the male and female will quickly part company.

The female solenodon only gives birth to one to two young every year, an astonishingly low number for a small mammal. She gives birth to her young in a subterranean nest in the burrow system she excavates with her powerful forelimbs. At first, the young are blind and naked, but they grow quickly, and before long, they are able to accompany their mother on her nocturnal forages. Amazingly, when baby solenodons accompany their mother, they never let go of her greatly elongated teats, so when the baby is really small, it simply gets dragged around, but as it grows, it is able to trot alongside its mother with the teat clasped firmly in its mouth. The young solenodon stays with its mother for several months, and even when it has ceased hanging on to her pendulous teats, it follows her around and licks at her mouth when she is feeding to learn the food preferences that will help it survive as a solitary adult.

As interesting as these insectivorous mammals are, they are completely defenseless against humans, and Marcano's solenodon has already been lost forever. The remains of this solenodon have been found with the bones of brown rats; therefore the species was still around when Europeans first reached Hispaniola as rats only reached the Greater Antilles aboard the ships of Columbus and later explorers. Amerindians reached Hispaniola thousands of years before Columbus arrived, and they appear to have had little effect on the populations of the Solenodon. As these animals are small and nocturnal, the first humans to settle Cuba probably only saw them rarely. The disaster for the solenodons, especially Marcano's solenodon, were the animals introduced by Europeans. Apart from occasionally falling victim to boa constrictors and raptors, solenodons had no enemies before the arrival of Europeans, and as a result, their defenses against cats and dogs are pitiful. If pursued by one of these predators, the solenodon stops in its tracks and hides its head between its forelimbs. Disastrously ill equipped to cope with the influx of new predators, Marcano's solenodon was wiped out, and the remaining Hispaniolan and Cuban species are now woefully endangered.

+ The solenodons are an ancient group of insectivorous mammals that have changed little in millions of years. They are known from North American fossils between 26 and 32 million years old.
+ There is some debate over the closest living relatives of the solenodons, but they are probably most closely related to the primitive tenrecs, another group of unusual insectivorous mammals found on the island of Madagascar and in parts of western and central Africa.
+ Apart from the solenodons, only a handful of other venomous mammals are known: the platypus (*Ornithorhyncus anatinus*), Eurasian water shrew (*Neomys fodiens*), short-tailed shrews of the genus *Blarina*, and slow loris (*Nycticebus coucang*). Why there should be so few venomous mammals is an interesting quandary, but it is probably because mammals have evolved a number of ways of catching their prey swiftly and efficiently. Even the best venom cannot bring about death immediately.

Further Reading: Woods, C. "Last Endemic Mammals in Hispaniola." *Oryx* 16 (1981): 146–52; MacFadden, B.J. "Rafting Mammals or Drifting Islands?: Biogeography of the Greater Antillian Insectivores *Nesophontes* and *Solenodon*." *Journal of Biogeography* 7 (1980): 11–22; Morgan, G.S., and C.A. Woods. "Extinction and Zoogeography of West Indian Land Mammals." *Biological Journal of the Linnean Society* 28 (1986): 167–203.

♀ Extinction Insight: Human Discovery and Extinction

In this chapter, you can read about some of the animals that have become extinct in the last 500 years or so. Many of these were birds, and many inhabited islands that only became known to Europeans during the last five centuries. Although the intensity of human movement started to really increase five centuries ago, migration and exploration are innate facets of human nature and are things we have always done. The search for food and companions and simple curiosity has driven us to look over the next hill or mountain range at what lies beyond. The ability to construct seafaring craft is likely a very ancient skill, and people have used boats and rafts to reach distant islands without knowing if there was any landmass to reach. This desire to move and explore is so ingrained in us that the bones of our ancestor *Homo erectus* have been found in Indonesia, thousands of miles from where the species originated—Africa. This was way before the age of trains, planes, and automobiles, and even horses, so our ancient ancestors dispersed largely by foot and, to a lesser extent, by seacraft. Modern man followed the same dispersal routes out of the ancestral homeland and eventually colonized the whole globe, apart from the poles.

The discovery of new lands was good for our species, but it has been incredibly bad for the animals with which we share the planet. Islands have been hit the hardest, particularly the ones that had been isolated long enough for their animal inhabitants to evolve traits suited to a predator-free environment such as flightlessness in birds. A huge number of islands were once home to flightless birds: Mauritius with the dodo, Madagascar with the elephant bird, and New Zealand with the moa. When humans discovered these islands, it was the beginning of the end for a wealth of species— animals that were perfectly adapted to their surroundings but powerless to resist humans— because of the animals that live with us and the habitat destruction we inevitably cause. The animals of larger landmasses were better placed to adapt to the human challenge as many of them could simply move into areas where humans had not reached. With this said, there is increasing evidence that human hunting and habitat destruction may have contributed to the extinction of the American and Australian megafauna. It is becoming increasingly clear that modern humans are the most destructive animals in earth's history.

Ever since humans started to spread around the globe, we have contributed to the rate of animal extinction, but this entered a new phase with the dawn of the new age: the era of discovery, when the wealthy courts of Europe funded expeditions using sailing ships in the hope of establishing trade routes and building empires. New lands were discovered every year for centuries, and this is the time during which animals like the dodo joined the roll call of extinction. Centuries later, in the eighteenth and nineteenth centuries, the age of discovery moved into yet another phase, and we started to ask more and more questions about the world around us. Scientific methods brought order and classification to the natural world, and the natural historians were born. They wanted to name, number, and collect the natural world's treasures, and every expedition to far-off lands was incomplete without a zoologist, botanist, or geologist. Initial reports of unusual creatures were met with skepticism by the scientific community, but as specimens began to trickle back to the learned institutions of Europe, scientists realized that the earth was home

Human Discovery and Extinction—Human exploration and discovery have been directly responsible for the extinction of many of the animals featured in this book. Few of these are better known than the dodo, a species that was wiped out in a little over 60 years. (Renata Cunha)

to a myriad of animal species, many of which were startlingly different to what they knew already. These new animals had to be collected and put on display. Live ones found their way into zoological gardens, and dead ones ended up stuffed or pickled in the museums that started to spring up all over Europe. This was an exciting time to be alive if you were a naturalist, but a very nervous one if you were an exotic, rare bird.

Museums and independently wealthy collectors would pay huge sums of money for specimens of rare animals. One of the most famous collectors was Lionel Walter Rothschild, a member of the Rothschild banking family, who devoted his life to the collection and study of nature. As a boy, he started off collecting butterflies, moths, and other insects, but he progressed on to larger animals, using his portion of the family fortune to secure rarities, especially birds. During his lifetime, Rothschild accumulated 2,000 mounted mammals, about 2,000 mounted birds, 2 million butterflies and moths, 300,000 bird skins, 144 giant tortoises, and 200,000 birds' eggs. He employed a small army of collectors to scour the far reaches of the globe for additions to his collections, and he was particularly keen to get his hands on species that had dwindled in numbers due to habitat destruction and human hunting and persecution. Rothschild was not alone as a fanatic collector of living things, and it is thought that together, these private collectors may have contributed to the extinction of several species, particularly birds that had already been pushed to the edge by human disturbance of their once pristine habitats.

Collecting still goes on today, and in some places, it is a real problem, but the tide of public opinion has turned against seeing stuffed animals in museums to appreciation of the living creatures in their natural environment. Sadly, the natural world is now confronted by the greatest man-made challenges: the spiraling population of our species and the wholesale destruction of habitats, both at a time when our understanding of the natural world has grown to a point where we can see the fragility of the world we live in and what we must do to save it.

Further Reading: Grayson, D.K. "The Archaeological Record of Human Impacts on Animal Populations." *Journal of World Prehistory* 15 (2001): 1–68; Grayson, D.K., and D.J. Meltzer. "Clovis Hunting and Large Mammal Extinction: A Critical Review of the Evidence." *Journal of World Prehistory* 16 (2002): 313–59; Grayson, D.K., and D.J. Meltzer. "A Requiem for North American Overkill." *Journal of Archaeological Science* 30 (2003): 585–93; Williams, J.R.S. "A Modern Earth Narrative: What Will Be the Fate of the Biosphere?" *Technology in Society* 22 (2000): 303–39.

4

FEWER THAN 10,000 YEARS AGO

MOA-NALO

Moa-Nalo—Many species of the giant, flightless ducks known as moa-nalo once inhabited the Hawaiian Islands. (Renata Cunha)

Scientific name: Several species
Scientific classification:
   Phylum: Chordata
   Class: Aves
   Order: Anseriformes
   Family: Anatidae
When did they become extinct? These birds became extinct around 1,000 years ago.
Where did they live? Their remains have been found on all the larger Hawaiian islands.

The island chain of Hawaii, located around 3,700 km from the U.S. mainland, is the most remote archipelago on the planet. The islands that make up Hawaii appeared from beneath the waves and are effectively the tops of submarine volcanoes that increase in height and area as they disgorge their very runny lava. Following their appearance, these landmasses were quickly colonized by living things. Bacteria, plants, fungi, and small animals can be dispersed on the wind, and the waves deposit other pioneers.

Birds, with their power of flight, are probably the first large animals to reach uncolonized islands, and one group of these animals, which reached Hawaii, evolved into bizarre creatures. These were the moa-nalo, and they were a group of flightless, gooselike birds that lived on all the main Hawaiian Islands. The word *moa* means "fowl" and *nalo* means "lost," so their Hawaiian name can be translated as "lost fowl." The remains of these birds have been found in sand dune blowouts, where the wind has uncovered their bones, and in sinkholes and lava tubes, both of which probably act as natural traps. These bones show that these birds were about the same weight as a swan, but much stockier, with a robust pelvis and powerful, thick legs. Moa-nalo also had very large bills that have been likened to the horny jaws of the giant tortoises that inhabit the Galápagos Islands and some of the islands in the Indian Ocean.

The moa-nalo may have been equipped with powerful bills and sturdy legs, but their wings were tiny structures that were of no use whatsoever for flight. Like the moa of New Zealand, the dodo of Mauritius, and the elephant bird of Madagascar, the moa-nalo had no need of flight as there were no large predators on the Hawaiian Islands. In this predator-free environment, the birds gave up flight and became large, ground-dwelling creatures.

What did these peculiar birds eat? The numerous remains that have been found of the moa-nalo include coprolites (fossilized droppings). These droppings have been studied, and it seems that the moa-nalo were specialist plant eaters. They probably waddled around the lush Hawaiian Islands nibbling a variety of low-growing plants. The beaks of some species of moa-nalo are even equipped with serrations that functioned like teeth, enabling them to take beakfuls of tough vegetation. The contents of plant cells are nutritious, but they are bound in a tough wall of cellulose that animals cannot digest because they lack the ability to produce the enzyme known as cellulase. To get at the goodness inside plant cells, any plant-feeding animal has to enlist the help of bacteria, and moa-nalo were no exception. Like horses and rabbits, the moa-nalo were hind-gut fermenters. The rear portion of their digestive tract was where the soup of mashed up plant matter and digestive fluids were brought into contact with the symbiotic, cellulase producing micro-organisms. More evidence for moa-nalo as plant eaters is the observation that many types of native Hawaiian plant are well protected with thorns and prickles. Such protection seems an extravagance on an island where there are no large native herbivores, but these defenses are probably reminders of the time when these plants were at the mercy of these plant-nibbling birds that roamed all over Hawaii.

Following the discovery of moa-nola remains, it was a mystery exactly what type of bird they were. In general size and proportion, they were gooselike, but the bones of the moa-nalo had more in common with ducks. Today, it is possible to extract DNA from long-dead bones and compare this to DNA taken from living species to build a family tree and to tell us how long a species has been around. Ancient DNA cannot give us 100 percent accurate results, but it can give us plausible estimates and scenarios. The DNA extracted from moa-nalo bones showed that these birds were indeed more closely related to the ducks and that their ancestor reached the Hawaiian Islands about 3.6 million years ago. What was their ancestor? It is difficult to know for sure, but some experts believe that the very widespread Pacific black duck (*Anas superciliosa*) or a now extinct similar species are likely candidates. The Hawaiian Islands, 3.6 million years ago, were a lush paradise without any large browsing animals, so the ancestors of the moa-nalo spread between the islands and evolved to fill this gap.

Like the numerous other flightless birds that have become extinct in the last couple of millennia, we can be almost certain that humans caused the extinction of the moa-nalo. The time of arrival of humans in Hawaii is a bone of contention among anthropologists, but Polynesians have been there since at least A.D. 800. Like the dodo, the moa-nalo was very easy to hunt. They had never seen a human and so had no innate fear of our very dangerous species. Moa-nalo were large birds (4 to 7 kg) and probably highly prized by Polynesian hunters. As the moa-nalo had evolved in the absence of predators, there was no need to reproduce quickly to balance out the mortality rate. They were probably very long-lived, slow-growing birds with a low rate of reproduction. The other big problem that humans brought with them to Hawaii was a menagerie of nonnative animals (dogs, cats, sheep, goats, pigs, etc.). These competed with the moa-nalo for food, disturbed their nests, and even ate their eggs. Even though they had lived, unmolested, on the Hawaiian Islands for more than 3 million years, the moa-nalo were probably hammered into extinction in as little as 200 years after the first humans reached this volcanic archipelago.

+ Hawaii is so distant from other landmasses that a huge variety of unique creatures evolved there. The birds were especially diverse, and a few ancestral colonists that reached these remote islands from distant shores gave rise to a myriad of species, many of which are now sadly extinct.

+ It is thought these original colonists were represented by 15 species, and over a short period of geological time, they evolved into around 78 species, although this number is far higher if we include those species, such as the moa-nalo, that are known only from bones.

+ Since humans colonized Hawaii, more than 56 species of bird have become extinct, and many of the remaining native species are severely endangered. The demise of some of these species is thought to have been caused by avian malaria, which was introduced to the islands by nonnative birds brought by humans.

Further Reading: James, H. F., and D. A. Burney. "The Diet and Ecology of Hawaii's Extinct Flightless Waterfowl: Evidence from Coprolites." *Biological Journal of the Linnaean Society* 62 (1997): 279–97; Sorenson, M. D., A. Cooper, E. E. Paxinos, T. W. Quinn, H. F. James, S. L. Olson, and R. C. Fleischer. "Relationships of the Extinct Moa-Nalos, Flightless Hawaiian Waterfowl, Based on Ancient DNA." *Proceedings of the Royal Society of London, B: Biological Sciences* 266 (1999): 2187–93; Slikas, B. "Hawaiian Birds: Lessons from a Rediscovered Avifauna." *Auk* 120 (2003): 953–60.

DU

Scientific name: *Sylviornis neocaledoniae*
Scientific classification:
    Phylum: Chordata
    Class: Aves
    Order: Galliformes
    Family: Sylviornithidae
When did it become extinct? The du is thought to have become extinct around 1,500 years ago, but it is possible that the species survived into more recent times.

Du—The 30-kg du constructed huge nest mounds on New Caledonia and the Île des Pins. (Renata Cunha)

Where did it live? The remains of this bird have been found in New Caledonia and the nearby island of Île des Pins.

In Australia, New Guinea, parts of Indonesia, and some of the Pacific islands live birds known by various names, including megapodes, brush-turkeys, mound builders, and incubator birds. These chicken-sized animals are unique among their feathered relatives for building large mounds, in which they incubate their eggs. The well-known malleefowl (*Leipoa ocellata*) of Australia scrabbles at the ground with its feet and beak to excavate a pit up to 3 m wide and 1 m deep. The male bird is actually responsible for digging, and he part fills the pit with leaf litter and other rotting vegetation before his mate lays her clutch of eggs into the waiting organic incubator. The male kicks soil into the pit and keeps on going until he has formed a big heap, which can sometimes be 0.6 m high and several meters across.

The mound of the malleefowl is quite an impressive structure for a small animal, so imagine the humps formed by a 30-kg, 1.5-m-tall extinct mound builder. On the Île des Pins, there are enormous, 4,000-year-old mounds, some 5 m tall and almost 50 m across, that were once thought to be burial mounds created by islanders. Excavations of these mounds revealed no human remains and no grave goods, leading to the theory they may have been built by a giant bird as incubator mounds. Four thousand years have passed since the mounds were first built, and in that time, the elements have probably eroded them, so they must have been considerably bigger when they were new.

Sadly, the du is not around today, and we can only guess at what this bizarre bird looked like in life. We have no idea what its closest relatives are, and it is not known if it was actually closely related to the living mound builders. With that said, it is often portrayed as a thickset animal, with a large bill and a bony lump above its eyes that was covered in a fleshy comb. Such a large, heavy bird was undoubtedly too big to take to the wing, and we can be quite confident that it was flightless like many other giant island birds. Along with what was an unusual outward appearance, the du had a number of skeletal peculiarities that set it apart from the majority of other birds. In most birds, the two collarbones are fused to form the bone that every meat eater knows: the wishbone. In birds, the wishbone strengthens the chest skeleton for the muscular forces that are generated during flapping flight. The du's collarbones were not fused. It has also been said that the rib cage and the pelvis have many similarities with those of dinosaurs.

With only fragmentary evidence available to us, we can only speculate on the way the du lived its life. The bird's skeleton does not carry any of the hallmarks of a formidable predator, so we can assume that it was probably a herbivore that may have extended its diet to include invertebrates. It may have used its powerful legs to scrape at the soil for nutritious roots and tubers, but we'll never know what food it ate and how it found it. Apart from the giant mounds on Île des Pins, the possible incubator mounds of the du, we have precious little information on the rest of its breeding behavior. Did several birds work collectively to build the huge mounds, or was each one the work of a single pair? Such large structures undoubtedly took a great deal of digging and subsequent back-filling, and the birds must have toiled day and night. It is possible that the mounds were built over time by generations of du. As these birds had given up the power of flight, New Caledonia and the Île des Pins must have been free of land predators, and therefore the mortality of the young birds must have been low. This scenario normally results in long-lived animals with very low reproductive rates, but in various places throughout these islands, there are abundant, fragmentary remains of the du, and it seems there were juvenile birds in profusion. This had led some experts to suggest that the du produced large clutches of up to 10 eggs, and if this was the case, the du's life span was probably fewer than 10 years, which is very low for such a large bird. Perhaps the birds were killed off by disease or intermittent harsh weather, forcing the populations to adapt and produce large numbers of young.

On its Pacific islands, the du probably lived a relatively peaceful existence, with no predators to worry about and only food and mating to concern its bird brain. This untroubled way of life was shattered by the arrival of humans, who reached these shores from the direction of Australia. It is thought that the first humans to reach these islands were from a diverse group of people known as the Lapita and that they probably made landfall on New Caledonia and the Île des Pins around 1500 B.C., but this date is debatable. As with other untouched islands around the world, the arrival of humans heralded death and destruction for the original inhabitants. A large, flightless bird like the du, with no innate fear of humans, was easy pickings, and its flesh would have been a welcome treat for seafarers who had probably eked out a survival on meager rations for many months. The nest mounds, with their sizeable clutches of big eggs, would also have been vulnerable to humans and their collected menagerie (dogs, pigs, rats, etc.), and nest raids hastened the decline of the du. It is thought that humans managed to wipe out the du about 1,500 years ago.

- It has been suggested that the du may have survived into more recent times as giant birds exist in the folklore of the present inhabitants of New Caledonia and the Île des Pins.
- New Zealand, New Caledonia, the Île des Pins, and surrounding islands in the western Pacific are the only visible parts of a great, submerged continent known as Zealandia, a landmass with an area greater than Greenland or India. Zealandia sank beneath the

waves around 23 million years ago. It once formed part of the giant landmass known as Gondwanaland, but all that we can see today are its highest reaches.

- ◆ The flora and fauna of New Caledonia are very special. Many of the plants and animals are endemic and relics of the flora and fauna that populated the now fragmented Gondwanaland. As there were no native New Caledonian mammals, the fauna was dominated by birds and reptiles, but along with the du, many of the other, large denizens of this unique place are sadly extinct.

Further Reading: Poplin, F., and C. Mourer-Chauviré. "*Sylviornis neocaledoniae* (Aves, Galliformes, Megapodiidae), oiseau Géant éteint de l'île des Pins (Nouvelle-Calédonie)." *Geobios* 18 (1985): 73–105; Steadman, D. W. "Extinction of Birds in Eastern Polynesia: A Review of the Record, and Comparisons with Other Pacific Island Groups." *Journal of Archaeological Science* 16 (1989): 177–205.

HORNED TURTLE

Horned Turtle—With their spiked heads and tails, the horned turtles are among the largest and most bizarre turtles ever to have lived. (Renata Cunha)

Scientific name: *Meiolania* sp.
Scientific classification:
    Phylum: Chordata
    Class: Sauropsida
    Order: Testudines
    Family: Meiolaniidae
When did it become extinct? The last of these turtles is thought to have become extinct about 2,000 years ago.
Where did it live? The bones of these extinct turtles have been found on Lord Howe Island, 600 km from mainland Australia and the islands of New Caledonia.

There would be very few people who would fail to recognize a turtle, such is the familiarity of these unusual reptiles. Although the fossil record is full of peculiar beasts, it has been said that the turtles are among the oddest vertebrates to have ever lived. Although their skeleton has the same bones as any other vertebrate, they are put together in a very different way. Their body is protected by a bony shell, which is, essentially, a hugely modified rib cage. The strength of this external carapace depends on the species, but it ranges from the leathery dome of the soft-shelled turtles to the almost impregnable shell of the giant tortoises. Also unique is the position of the hip and shoulder girdles, as they are found inside the rib cage. These animals are most familiar for being able to withdraw their heads and legs into the safe confines of their shells. The way they withdraw their head allows scientists to identify two groups of turtle: the cryptodires and the pleurodires. The latter are often called side-necked turtles because they bend their long necks into an S shape to keep their heads out of harm's way. The turtles that people often keep as pets fall in the first group, the cryptodires, and these can pull their heads right into their shells by bending their necks below the spine.

There's no doubt that some of the turtles, especially the land-dwelling species, are very slow, lumbering creatures, characteristics that are often linked to evolutionary failure and poor adaptability. However, nothing could be father from the truth for the turtles. These shelled reptiles are a successful group of animals that have been around since the Triassic—at least 215 million years (and probably considerably longer)—which makes them much older than the lizards and snakes. Not only are they ancient, but they are among the very few living reptiles that have become almost completely amphibious, only leaving the water to lay eggs (some species of snake also only leave the water to lay eggs). Today, there are around 300 turtle species, ranging from tiny, 8-cm tortoises all the way up to the oceangoing giant, the leatherback turtle (*Dermochelys coriacea*), which can be 3 m long and weigh 900 kg.

Even though some truly bizarre turtles are still with us today, they pale in insignificance compared to an immense, land-living turtle that only became extinct in the last couple thousand years. This was the horned turtle, and in life it must have been an astonishing animal. The horned turtle was around 2.5 m long, and it must have weighed in the region of 500 to 700 kg. By comparison, the largest living land-dwelling turtle is the Galápagos tortoise (*Geochelone nigra*) at about 300 kg and 1.2 m long. Imagine a horned turtle alongside a Galápagos tortoise and you get an idea of the size of this extinct beast. Not only was the horned turtle big, but it also had a very bizarre appearance. Sprouting from its skull were large horns and spikes, the longest of which grew from toward the back of the head and could reach a span of 60 cm. This formidable forward armory was combined with the typical tortoise carapace and a heavily protected tail that also sported spines. The horns of this extinct turtle made it impossible for the head to be pulled into the shell during times of danger. It is possible that these horns were used by the turtle to defend itself, but we don't know what predators lurked on the islands where these extinct reptiles lived. Male giant turtles can be quite aggressive to one another during the breeding season, and maybe the extinct giant used its horns and tail spikes to fight other males for the right to mate. As with other island animals, the horned turtles may have grown to great size because there was very little in the way of threats in their isolated home terrain. Alternatively, great size is a simple yet effective defense against many predators. The truth is that we'll never know the evolutionary force behind the incredible size and appearance of these turtles.

What we can be more sure of is their diet. Large land-dwelling turtles are slow, heavy animals, so fast-moving animal prey is out of the question. We know that the Galápagos tortoise and other terrestrial giant turtles are herbivores that eat a wide range of plant matter. The horned turtle was obviously unsuited to climbing trees or rearing up on its back legs to reach lofty vegetation, so it must have been dependant on the unique, low-growing plants that grow on New Caledonia and the surrounding islands. All living turtles lay eggs, and we can assume that the horned turtle was no different, but how it laid them and where will never be known for certain. Perhaps it excavated a pit before laying its eggs and forgetting about them.

It is amazing to think that these giant, bizarre turtles roamed some of the isolated islands of the western Pacific into very recent geological times, but exactly why they died out is another mystery. We do know that island animals have suffered badly at the hands of humans, and we can be almost certain that the first thing to spring to the mind of the first human who saw these shelled giants was, "Can I eat it?" A slow-moving turtle, regardless of its size, is no match for humans and their various weapons. Lord Howe Island and New Caledonia are small areas of land, and they could never have supported large populations of such big animals; therefore it is very likely that when humans did discover the horned turtle, they wiped them out in a matter of centuries, or possibly even decades.

- Apart from the way that living turtles bend their necks to hide their heads, we can divide them another way into three groups: there are marine forms, with legs modified into flippers, for example, the leatherback turtle; terrestrial forms, with thick, pillarlike legs, for example, the Galápagos tortoises; and semiaquatic forms, for example, terrapins and snapping turtles.

- Many of the living species of turtle may soon follow the horned giant to extinction as they are incredibly endangered. Some of the very rare species only survive in small populations on isolated islands, while the oceangoing species are at risk from fishing hooks, drift nets, and direct hunting. Without complete and active protection, it is very likely that some of the most amazing turtles could be extinct within 30 years.

- As turtles lead such slow lives, they are among the most long-lived of the all the vertebrates. The Galápagos tortoise can live to be at least 150 years old. One famous, long-lived radiated tortoise (*Geochelone radiate*) was presented to the Tongan royal family in 1777 by none other than Captain Cook. Known as Tu'i Malila, this tortoise died in 1965, at age 188. The longevity of an immense turtle like the horned giant can only be guessed.

- Further back in the fossil record, in the age of the dinosaurs, there were other extinct turtles that were truly enormous. One of these, *Archelon*, is only known from 70-million-year-old fossils. It was about 4 m long, and the span of its flippers was around 4.5 m. Fully grown, *Archelon* probably weighed in the region of 2 to 3 tonnes. Its large head and powerful bite appear to be suited to eating shelled mollusks such as the extinct ammonites.

Further Reading: Gaffney, E.S. "The Postcranial Morphology of *Meiolania platyceps* and a Review of the Meiolaniidae." *Bulletin of the American Museum of Natural History* 229 (1996): 1–166; Gaffney, E., S. Hutchison, J. Howard, F.A. Jenkins, and L.J. Meeker. "Modern Turtle Origins: The Oldest Known Cryptodire." *Science* 237 (1987): 289–91.

GIANT LEMUR

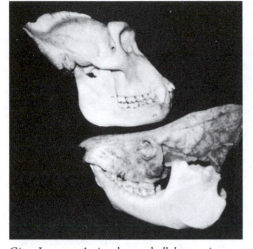

Giant Lemur—A giant lemur skull, *bottom*, is compared with a gorilla skull, *top*, giving an idea of how large this extinct Madagascan primate was. (Elwyn L. Simons)

Giant Lemur—Madagascar was once home to a number of very large lemurs. The skulls of some of these are shown in this photograph alongside two living species. *Above left to right: Megaladapis* (giant lemur), *Archaeoindris, Paleopropithecus* (sloth lemur), and *Archaeolemur* (all extinct). *Below left to right: Hadropithecus* (extinct) and the living smallest and largest lemurs, *Microcebus* and indri, respectively. (Alison Jolly)

Scientific name: *Megaladapis edwardsi*
Scientific classification:
  Phylum: Chordata
  Class: Mammalia
  Order: Primates
  Family: Lepilemuridae
When did it become extinct? The giant lemur became extinct around 500 years ago, perhaps even more recently.
Where did it live? The giant lemur was found only in Madagascar.

Many, many millions of years ago, what we know today as Madagascar was part of Gondwanaland, the enormous landmass that occupied the Southern Hemisphere. Madagascar was hemmed in by Africa to the west and India to the east, but over the ages, the slow but ceaseless movements of the immense plates that make up the surface of the earth tore Gondwanaland apart, and around 165 million years ago, Madagascar drifted free of Africa, but over the next 40 million years or so, it still retained intermittent contact with India. It lost touch with India for the last time around 88 million years ago, and ever since, it has been isolated in time and space. It is this isolation that makes Madagascar such an interesting place from a biological point of view. Around 75 percent of the larger Madagascan animals are found nowhere else on earth

The ancestors of some of the animals and plants that inhabit Madagascar were marooned as the island became more and more isolated, but the lemurs, probably the most familiar of all Madagascan animals, are thought to have evolved from an ancestor that inadvertently reached the island from Africa by drifting on a raft of floating vegetation.

There were once around 50 species of lemur living on Madagascar, but tragically, 15 or more species have become extinct since humans arrived on the island. It is possible that all of these lemurs evolved from a single ancestral species that floated across from Africa.

Essentially, the lemurs are primates, albeit primitive ones, and all of their close relatives that once lived in other parts of the world have long since become extinct, probably outcompeted by the ancestors of the Old World monkeys and apes. However, the lemurs were safe from competition on Madagascar, and there they flourished, evolving into a variety of forms to exploit the various habitats on the huge island. Today, the pygmy mouse lemur (*Microcebus myoxinus*) is the smallest living lemur at around 30 g, whereas the largest, the indri (*Indri indri*), can weigh as much as 10 kg. Like any other group of animals, the lemurs were not without their giants, and up until 500 years ago, Madagascar was home to some enormous lemurs.

Lots of skeletons and individual bones of the giant lemur have been unearthed from sites on the west coast of Madagascar, and they belong to an animal with bodily proportions comparable to a koala bear. The fingers and toes of the giant lemur were very long indeed and probably enabled the living animal to get a good grip on tree trunks. Like the living koala bear, the giant lemur probably spent the majority of its time in the trees. The jaws and the teeth of this primate are very robust, and it probably used them to good effect to chew leaves. The giant lemur's canines are well developed, and it probably used these during the breeding season, when disputes over territory and mates broke out, as well as for protecting itself from predators, however, this primitive primate lacked upper incisors. Projecting from the nose of the giant lemur's skull is a bony lump, very similar to the structure that can be seen on the skull of a black rhinoceros, and like this large ungulate, the giant lemur may have had a prehensile upper lip to bring leaves to its mouth.

This extinct lemur was undoubtedly equipped to defend itself, but from what? The largest mammalian predator found in Madagascar today is the fossa (*Cryptoprocta ferox*), a very agile animal whose closest living relatives are the mongooses. However, at 10 kg, a fossa was no match for this large, powerful primate. Recent finds show that Madagascar was once home to a giant fossa (*Cryptoprocta spelea*), a predator that was about 1.8 m long and 17 kg in weight, and like the living fossa, this giant was nimble and at home in the trees. It is this animal that probably preyed on the giant lemur.

Apart from the giant fossa, the giant lemur had nothing to fear, that is, until the arrival of humans. The story of the colonization of Madagascar by humans is an interesting one. Sometime between A.D. 200 and 500 (about the same time as England was being colonized by the Saxons), seafarers from Borneo set off across the great expanse of the Indian Ocean without any knowledge of what was before them. After traveling counterclockwise around the Indian Ocean, a distance of almost 6,000 km, without compasses or charts, they reached Madagascar. This was a massive achievement for them but a disaster for the amazing wildlife of this island. These first human inhabitants brought animals and agriculture, and the landscape and wildlife of Madagascar was changed, irrevocably, for the worse. A 50-kg animal like the giant lemur must have been prized as food, and as the forests were cleared to make way for crops, the native animals of the island were squeezed into smaller and smaller patches of habitat. Shortly after the arrival of the Indonesians, Bantu people from the east coast of Africa also migrated to Madagascar, and they brought their own types of devastation.

The giant lemur probably clung to existence until around 500 years ago, and it was almost certainly still in existence when the Portuguese first reached this island in A.D. 1500. Interestingly, it appears that the Malagasy people were terrified of the giant lemur species and would apparently run away in fear whenever they chanced on one. After generations of persecution, the feeling was probably mutual, and the giant lemurs probably did everything they could to keep out of the way of humans, until the forests had dwindled to such an extent that there was nowhere left to hide.

- To say that Madagascar has been trashed is an understatement. Since humans colonized the island, around 90 percent of the original forest cover has been lost. This treasure trove of biological diversity has been reduced to a shadow of its former glory. Indeed, we only have a rough idea of how many species of unique animal and plant have disappeared since humans first arrived.
- The giant lemur was not the only large lemur to once live in the forests of Madagascar. Another extinct species, *Archaeoindris fontoynonti*, may have been the size of a gorilla, while other species, such as *Palaeopropithecus* sp., slightly smaller than the giant lemur, lived a more sedentary lifestyle and are known as sloth lemurs.
- In Malagasy folklore, there are tales of the animal known as the *tretretretre*. In 1661, the French explorer Etienne de Flacourt made many observations on the natural history of Madagascar, including this account of the *tretretretre* from his 1661 tome, *L'Histoire de le Grand Île de Madagascar*: "The tretretretre is a large animal, like a calf of two years, with a round head and the face of a man. The forefeet are like those of an ape, as are the hindfeet. It has curly hair, a short tail, and ears like a man's. . . . It is a very solitary animal; the people of the country hold it in great fear and flee from it, as it does from them." It is highly likely that these tales relate to the sloth lemurs.

Further Reading: Fleagle, J. G. *Primate Adaptation and Evolution*. New York: Academic Press, 1988.

WOOLLY MAMMOTH

Woolly Mammoth—A herd of woolly mammoth wondering across the steppe must have been an imposing sight. (Phil Miller)

Scientific name: *Mammuthus primigenius*
Scientific classification:
    Phylum: Chordata
    Class: Mammalia
    Order: Proboscidea
    Family: Elephantidae
When did it become extinct? The woolly mammoth is thought to have become extinct around 10,000 to 12,000 years ago, although a dwarf race of this species survived until around 1700 B.C.
Where did it live? The woolly mammoth roamed over a huge area of the prehistoric earth, including northern North America and northern Eurasia.

What African safari would be complete without a sighting of an elephant? We associate these majestic animals, the largest of all land-living animals, with warm places, yet thousands of years ago, the world was a very different place—a much chillier place—and a long-dead relative of the elephants we know today actually thrived in bitterly cold conditions. The species was the woolly mammoth, and in essence, it was an elephant covered in a dense pelage of shaggy hair.

A fully grown woolly mammoth was around 3 m tall at the shoulder and probably weighed in the region of 7 tonnes, which is quite a lot smaller than a large African bull elephant (3.5 m tall and 10 tonnes in weight), but its dense fur made it look very imposing. The remains of the woolly mammoth have been found in many locations, and some of them are in excellent condition, which allows us to build a very good picture of what the living animal was like. We know that the dense fur of the mammoth was around 50 cm long, and we also know what color this fur was—some of these huge beasts had dark brown fur, while others had pale ginger or even blonde fur. The fur of the woolly mammoth, coupled with an 8-cm layer of fat beneath the skin, served as insulation from the terrible cold of the ice age tundra. Sebaceous glands in the skin of the mammoth exuded greasy oil into the shaggy coat to enhance its insulating properties. Another interesting adaptation protected them from the cold still further: a patch of hairy skin that hung over the anus to prevent the escape of precious warmth. The African elephants are renowned for their ears, which in large specimens can be around 1.8 m long, but the woolly mammoth's ears were only around 30 cm long—yet another adaptation to a cold climate as a greater surface area of skin will allow more of the body's heat to escape. Large ears help an African elephant to stay cool, but the mammoth was struggling to stay warm.

Apart from its shaggy fur, the other striking feature of the woolly mammoth was its enormous, curving tusks. The tusks of elephants are actually teeth that have grown out of the mouth, and in the woolly mammoth, they kept on growing until they were around 4 m long. Like in modern elephants, these tusks were probably important to establish a pecking order among the males when it came to the breeding season, which may well have been at the end of July and the beginning of August. Tusks are a measure of the owner's strength, and they can be flaunted to assert dominance without the need for fighting and the potential injury it may bring; however, when two evenly matched males came head to head, a fight was probably inevitable. The front of the mammoth's head was quite flat; therefore males could have butted heads and locked tusks. Using all of their strength, the male mammoths wrestled with the intent of digging the tusks into the flanks of their opponent.

As the woolly mammoth is no longer alive, we can only make assumptions about the way it lived, but it is highly likely that it formed family groups like those formed by the African and Indian elephants—close-knit groups that are led by a female and comprise adult females and young. Like elephants, mature male woolly mammoths probably banded together in loose groups until the breeding season arrived, when they searched out the female-led groups. As in elephants, pregnancy in mammoths probably lasted around 22 months, with a single infant being born at the end of that time.

The remains of the woolly mammoth that have been found even tell us what this animal ate. The tundra where the woolly mammoth lived was devoid of large trees, and these huge animals probably relied on coarse grasses and low-growing shrubs, such as dwarf birch and willow, for sustenance. As tundra vegetation is far from the most nutritious plant matter, it is reasonable to assume that the woolly mammoth needed to consume huge quantities of this tough vegetation to sustain its great bulk.

The woolly mammoth was around for at least 290,000 years; however, its reign ended at the end of the last glaciation, which in geological terms was quite abrupt, but as the mammoth had survived numerous cycles of climate change, where long glaciations have been interspersed with shorter, warmer intervals, something else must have been happening. It has been observed that the disappearance of many of the world's large land-living animals at the end of the last ice age coincides with the dispersal of humans north from more temperate latitudes and into the New World. As the ice age relaxed its grip, humans edged farther and farther north into areas that had previously been inhospitable, and we know that these prehistoric people, our ancestors, hunted the mammoth for its meat and all the other parts of its body, which their skilled hands could turn into clothes, tools, and shelters. It is very possible that the human species contributed to the extinction of many majestic animals, including the woolly mammoth.

+ Ten species of mammoth have been identified from around the world, and the group is thought to have evolved from an ancestor that lived in North Africa about 5 million years ago.
+ The woolly mammoth was not nearly as large as some of the other mammoth species. The steppe mammoth (*Mammuthus trogontherii*), the Columbian mammoth (*Mammuthus columbi*), and the imperial mammoth (*Mammuthus imperator*) were all very large, and the latter species could have measured 5 m at the shoulder and weighed in excess of 13 tonnes. The Songhua River mammoth (*Mammuthus sungari*) may be one of the largest terrestrial mammals ever, at 17 tonnes.
+ A population of dwarf woolly mammoths survived on Wrangel Island in the Arctic Ocean north of Siberia for a long time after the rest of the species went extinct—possibly as recently as 1700 B.C. Other island populations of dwarf mammoths existed on Sardinia and the islands off the coast of California.
+ Woolly mammoths are almost unique among the prehistoric fauna for their incredibly well-preserved remains. Numerous specimens—adults and young—have been found in the permafrost of what is now Siberia. The most recent find was a perfectly

preserved body of a 10,000-year-old female mammoth calf, found near the Yuribei River in Russia. To date, 39 preserved woolly mammoths have been found, but only four of these are complete. A trade still exists today in the ivory tusks from these long-dead animals.

+ Scientists have speculated that it would be possible to bring the mammoth back to life using the technology of cloning and the tissue from the mammoths that have been frozen in permafrost. This is an interesting notion, but the flesh of the frozen specimens, even when newly discovered, is badly decayed, and the DNA is unsuitable for cloning.

Further Reading: Guthrie, R.D. "Reconstructions of Woolly Mammoth Life History." In *The World of Elephants—International Congress, Rome*, 276–79. 2001; Gee, H. "Evolution: Memories of Mammoths." *Nature* 9 (2006): 439; Solow, A.R., D.L. Roberts, and K.M. Robbirt. "On the Pleistocene Extinctions of Alaskan Mammoths and Horses." *Proceedings of the National Academy of Sciences USA* 103 (2006): 7351–53; Lister, A.M., and A.V. Sher. "The Origin and Evolution of the Wooly Mammoth." *Science* 294 (2001): 1094–97.

SIVATHERE

Scientific name: *Sivatherium* sp.
Scientific classification:
    Phylum: Chordata
    Class: Mammalia
    Order: Artiodactyla
    Family: Giraffidae
When did it become extinct? Estimations for the disappearance of these animals vary, but there is a slight possibility that a species of *Sivatherium* may have survived until as recently as 5,000 years ago.
Where did it live? The bones of sivatheres have been found throughout Asia, Europe, and Africa.

In the 1930s, a joint expedition of the Field Museum in Chicago and Oxford University carried out some excavations of an early Sumerian site in Kish, Iraq. One of their finds was a copper rein ring, designed to fit on the tongue of a chariot. Decorating the top of this ring is an unusual, horned ungulate. Sumerians normally decorated such pieces with sculptures of horses, but the animal depicted in the Kish rein ring is unlike any living animal. Archeologists, without any knowledge of long-dead beasts, described the mystery animal as a stag, but a young paleontologist saw this ring and realized at once that the sculpture surmounting it appeared to be a *Sivatherium*, a relative of the giraffe that was previously thought to have disappeared around 1 million years ago. The horns of the sculpture have been broken, but Edwin Colbert recognized the distinctive shape from the *Sivatherium* skulls he had seen. More important, he noticed the stumpy horns between the eyes and the large horns of the sculpture, a feature unique to *Sivatherium*. Is it possible that this unmistakable animal survived into recorded history and was known to the early Sumerians?

Sivathere—There's a possibility that a *sivathere* survived into the era of recorded history. Note the large horns and bony ossicones above the eyes. (Phil Miller)

Sumerian craftsmen were not prone to flights of fancy, and the animals we see in Sumerian rein rings are clearly real species known to these people. The animal in the Kish artifact also has a long, heavy rope extending from its snout. Could this indicate that it had been captured alive and tamed?

If this theory is proven to be correct, it is an amazing discovery. *Sivatherium* belongs to a group of animals collectively known as sivatheres, the largest of which was around 2.2 m at the shoulder. Like the living giraffe (*Giraffa camelopardalis*) and the shy, forest-dwelling okapi (*Okapia johnstoni*), *Sivatherium* was a herbivore feeding on grasses, leaves, or bushes, very much like the moose (*Alces alces*). Offshoots of the giraffe family, the notable characteristics of the sivatheres were the huge horns adorning their strengthened skulls and a pair of stubby horns (ossicones) above their eyes. The combination of a strong skull and huge horns made the head very heavy, and the neck and shoulders of the animal were very strong to support the great weight. The powerful forequarters of *Sivatherium* were taller than the hindquarters, giving the animal a sloped back. Modern giraffes have horns, too, but they are relatively short and covered in skin, and the males of these long-necked animals use these horns during the breeding season to assert their dominance over their rivals by standing side by side and swinging their heads into the flanks of their opponents. *Sivatherium* must have also used its horns during the breeding season, but the large horns may have simply been for display. When two evenly matched males came face-to-face, they probably butted heads and wrestled with their ossicones locked together.

Sivatheres appear to have evolved in Asia around 12 million years ago (mid-Miocene). Miocene sivathere fossils are also known from Iran, Turkey, Greece, Italy, and Spain. The animals probably immigrated into Africa at a later date as the oldest sivathere remains from this continent are 5 million years old. These unusual giraffids were successful animals that diversified into several species that succeeded in colonizing a huge area of the ancient earth. What happened to them? Before Edwin Colbert made the link between the Kish artifact and *Sivatherium*, this extinct beast was thought to have disappeared around 1 million years ago, the victim of climatic change and competition from other herbivorous ungulates. If the Kish artifact depicts a genuine animal that survived until at least 3500 B.C., our explanations for the disappearance of *Sivatherium* are inaccurate. The discovery of this small copper sculpture has provided us with the intriguing possibility that a species of sivathere survived until the cusp of recorded history and actually occupied a place in the traditions and customs of the Sumerian people. When the Sumerian metalworker created the copper sculpture surmounting the rein ring, *Sivatherium* may have been clinging to survival in the remote reaches of modern-day Iraq. This fascinating story presents us with the possibility that many other extinct animals survived into far more recent times than bones alone suggest.

+ The rock paintings of Tassili n'Ajjer, in the Algerian Sahara, depict many different animals, including what appears to be an unusual giraffid. Is this yet more evidence for the survival of *sivatheres* into relatively recent times?

+ The surviving relatives of *Sivatherium* are the giraffe and the okapi, both of which are only found in Africa. The okapi is a shy, forest-dwelling animal that only became known to science in 1912.

+ Five thousand years ago, Kish in Iraq, was a very different place. It sat at the eastern edge of what has become known as the Fertile Crescent, the arc of land watered by three enormous rivers: the Nile, Euphrates, and Tigris. Today, these areas are semiarid, but five millennia ago, rainfall was much higher. The whole area was very productive and it is considered to be the cradle of civilization, where people first turned from a hunter-gatherer existence to settled societies underpinned by agriculture. These verdant valleys may have been the last stronghold of the sivatheres, and such a rare, impressive animal would have undoubtedly been held in high regard by the earliest civilizations.

Further Reading: Colbert, E. H. "Was the Extinct Giraffe (Sivatherium) Known to the Early Sumerians?" *American Anthropologist* 38 (1936): 605–8.

GIANT DEER

Giant Deer—The giant deer was about the same size as the moose, but its antlers were enormous. Some are more than 3.6 m across. (Renata Cunha)

Scientific name: *Megaloceros giganteus*
Scientific classification:
    Phylum: Chordata
    Class: Mammalia
    Order: Artiodactyla
    Family: Cervidae
When did it become extinct? The most recent remains yet recovered of the giant deer
    have been aged at around 7,000 years using radiocarbon dating, but the species could
    have survived into more recent times.
Where did it live? This deer was a found throughout Europe and east into central Asia.

Several well-preserved skeletons of this splendid beast have been found in the peat bogs of Ireland, which is why it used to be familiarly known as the "Irish elk"; however, its range was not restricted to Ireland. Bones of this animal have turned up all over great swathes of Europe, and more rarely, in Asia, and it was certainly known by our forebears.

The giant deer was another species from the group of mammals collectively known as megafauna. Like many of the animals that adapted to the cold conditions of the ice age, the giant deer grew to a great size. In stature, it was a little larger than an average moose (*Alces alces*), measuring about 2.1 m at the shoulder. This is impressive enough, but the antlers of the male were enormous. Skeletons have been found with antlers weighing 40 kg, which have a span of more than 3.6 m. Why this deer should have such huge adornments on its head has been a source of heated debate for some time, but it is now generally accepted that like the majority of extravagant male adornments found in nature, the antlers were a product of sexual selection and no bigger than expected when we take into account the size of the animal that carries them. In deer and their relatives, the size and structure of their antlers is important when it comes to the breeding season. A male deer's antlers are a measure of how strong and fit he is. In many cases, two male deer do not have to fight to work out who is the more dominant as simply posturing and showing off the antlers will suffice, but when the need arises, they are potent weapons, and stags will lock antlers, wrestle, and attempt to injure one another.

By looking at the size, structure, and placement of the giant deer's antlers on its head and the structure of the animal's skull, it is very likely that males of this species fought, especially when two equally matched stags crossed paths. The bone at the top of the skull was also very thick (3 cm), a necessary reinforcement if the head was not to be sheared in two by the forces exerted during a fight. It has been suggested that after some bellowing and posturing, a pair of well-matched stags lowered their heads in between their front legs and locked antlers. Using all their body weight, they tried to inflict wounds on the flanks of their opponent. A pair of fighting giant deer stags straining and kicking up clouds of dust must have been a magnificent sight.

It is likely that, as with other deer, the antlers of the male giant deer were shed annually. Growing such enormous structures from the top of the head must have placed great stress on the male, who must have had to increase his food consumption considerably to fuel the growth of the gigantic structures. The giant deer's dietary requirements were probably very similar to modern deer, feeding mainly on grasses. It is also possible their great size allowed them access to high-growing vegetation that was out of reach for other deer and browsing mammals.

+ The span of the giant deer's antlers was a severe handicap in heavily wooded habitats, so we can assume that it was an animal of open country, where it could find abundant food. These open spaces would also have provided the giant deer with some degree of protection from its enemies as predators must have found it difficult to take the deer by surprise.

+ The National Museum of Ireland has more than 200 specimens of giant deer skulls and antlers, all of which were found in the country's peat bogs and lakes. Peat bogs are excellent preservers of ancient remains as there is very little oxygen present for the bacteria that are responsible for the process of decay. Well-preserved bones and tissue, thousands of years old, can be found in the lake clays, with only peat staining to show for their long entombment.

+ Most of the Irish specimens have been found beneath the peat in a layer known as lake clays. Geologists know that these clays were deposited between 10,600 and 12,100 years ago and belong to a period of time known as the Woodgrange Interstadial. This period occurred toward the end of the last ice age and was marked by a climate that was similar to today's. This period produced conditions perfect for preservation, which is why we find so many specimens of giant deer from this time.

+ The most complete skeleton of a giant deer, now on display at the Paleontological Institute in Moscow, was discovered near the Russian town of Sapozhka. This fine specimen really gives a sense of how imposing the living animal must have been.

Further Reading: Lister, A.M. "The Evolution of the Giant Deer, *Megaloceros giganteus* (Blumenbach)." *Zoological Journal of the Linnean Society* 112 (1994): 65–100; Moen, R.A., J. Pastor, and Y. Cohen. "Antler Growth and Extinction of Irish Elk." *Evolutionary Ecology Research* 1 (1999): 235–49.

GIANT GROUND SLOTH

Giant Ground Sloth—The human silhouette in this picture gives an idea of how huge these extinct sloths were. They could even rear up on their hind legs to reach lofty food. (Natural History Museum at Tring)

Giant Ground Sloth—The ground sloths were perhaps the most impressive of all the extinct South American mammals. The largest species (*Megatherium americanum*), the one depicted here, was about the same size as a fully grown elephant. (Renata Cunha)

Scientific name: *Megatherium americanum*
Scientific classification:
    Phylum: Chordata
    Class: Mammalia
    Order: Pilosa
    Family: Megatheriidae
When did it become extinct? The last giant ground sloths are thought to have died out around 8,000 to 10,000 years ago.
Where did it live? The giant ground sloths were found throughout South America.

South America is probably the most biodiverse landmass on earth, yet, many thousands of years ago, the fauna of this continent was even more remarkable. A perfect example of this long-gone South American fauna is a ground-dwelling sloth that was the same size as an elephant. This was the giant ground sloth, and it was an immense and unusual animal. Fully grown, the giant ground sloth was about 6 m long, and estimates of its weight range between 4 and 5 tonnes. Several skeletons (real and copies) of this animal are to be found in museum collections around the world, and one of the most astonishing things about these remains is the size of the bones. The limb bones and their supporting structures are massive and give an impression of a heavy, powerful animal. In life, the digits of the animal were tipped with long claws, which may have been used to grab plant food or even as weapons.

We know from the skeletons of this animal that the bones of the hind feet were arranged in a very peculiar way, making it impossible for the living animal to place its feet flat on the ground. The animal could certainly rear up onto its hind legs, and perhaps even manage to amble around in this posture, using its thick tail as a strong prop, but it had to shuffle around on the outside of its feet with the long claws pointing inward. The giant ground sloth may have been able to make better progress on all fours, possibly reserving its two-legged stance for feeding or defense.

As the giant ground sloth is related to the living sloths, it was always assumed that they were gentle plant-eating animals, but some recent, controversial scientific research has shed some light on how this massive beast used its forelimbs. These studies suggest the forelimbs of a giant ground sloth were adapted for fast movement. Such an ability was of little use to a plant-nibbling animal that needed a strong, sustained pull to bring tasty leaf-bearing branches within reach of its mouth. The research suggest that the muscles of the forelimbs were used to power the large claws into other animals, and maybe not only in defense. The animal's teeth also give intriguing insights into the way it fed. They are not the normal grinding blocks that are found in the mouths of plant-feeding mammals. They and the jaws they sit in appear to be adapted for slicing, much like the jaws and teeth of meat-eating animals. The claws and teeth of this giant mammal have led some people to suggest that the giant ground sloth was not a plant feeder at all, but a scavenging animal that used its size to drive predatory animals from their kill before digging in to the carcass. The image of a 5-tonne brute ambling over to a group of dire wolves, scaring them off, and then devouring their kill is quite fantastic. Regardless of this research, it is decidedly unlikely that this giant lived in this way, and like its living relatives, the giant ground sloth was probably a herbivore, but it may have been able to use its forelimbs and teeth to defend itself.

As with almost all of the long-dead animals that once roamed South America, we cannot be certain what brought about the demise of the giant ground sloth. It has been speculated that the arrival of modern humans, with spears and arrows, led to their extinction, but it is reasonable to assume that there was something much more far-reaching happening at the time that wiped these animals out. Climate change is one of the usual suspects, and we know that the earth's habitats were going through some massive changes at the time these animals went extinct. Global temperatures were changing, and land-dwelling animals everywhere were being affected. Hunting may have had an effect, but it may have been minor compared to the ravages of climate change.

Today, there are still vast areas of South America where people rarely venture, and some people believe that a species of giant ground sloth may have somehow survived the events that wiped out its relatives and is alive and well in these remote areas. Local inhabitants call the beast the *mapinguary*, and it is said to rear up on its back legs and emit a foul-smelling odor from a gland in its abdomen—not only that, but the creature is said to be impervious to bullets and arrows, thanks to some very tough skin on its belly and back. Without a specimen or an excellent photograph, it is difficult to take these stories seriously, but it is worth remembering that previously unknown species of mammal are discovered fairly regularly, and some of them are surprisingly large. If a live giant ground sloth was found today, it would be the zoological story of all time.

+ It is thought that there were around four species of giant ground sloth. The species mentioned here (*Megatherium americanum*) was by far the biggest. The closest living relatives of these extinct animals are the anteaters, armadillos, and tree sloths. The biggest of these, the giant anteater, would be dwarfed by even the smallest giant ground sloth.

+ In 1895, a rancher by the name of Eberhardt found some hide in a cave in Patagonia that turned out to be giant ground sloth skin. The skin was in very good condition, and some people believed that it was from an animal that died relatively recently. When techniques became available to age the skin, it was found to be several thousand years old—it was just that the very dry conditions in the cave had prevented it from rotting. Interestingly, the mummified skin was studded with bony nodules, which probably gave the animal excellent protection from the teeth and claws of predators, and perhaps even the spears and arrows of early humans.

+ It would be fantastic if a species of giant ground sloth had somehow survived into the modern day, but accounts of the *mapinguary* may be due to confusion with other animals or derived from folk memories of when humans encountered these animals thousands of years ago.

Further Reading: Bargo, M.S., G. De Iuliis, and S.F. Vizcaíno. "Hypsodonty in Pleistocene Ground Sloths." *Acta Palaeontologica Polonica* 51 (2006): 53–61; Bargo, M.S., N. Toledo, and S.F. Vizcaíno. "Muzzle of South American Pleistocene Ground Sloths (Xenarthra, Tardigrada)." *Journal of Morphology* 267 (2006): 248–63; Bargo, M.S. "The Ground Sloth *Megatherium americanum*: Skull Shape, Bite Forces, and Diet." *Acta Palaeontologica Polonica* 46 (2001): 173–92; Fariña, R.A., and R.E. Blanco. "*Megatherium*, the stabber." *Proceedings of the Royal Society of London* 263 (1996): 1725–29.

CUBAN GIANT OWL

Scientific name: *Ornimegalonyx oteroi*
Scientific classification:
    Phylum: Chordata
    Class: Aves
    Order: Strigiformes
    Family: Strigidae
When did it become extinct? It is thought that this giant owl became extinct around
    8,000 years ago.
Where did it live? The remains of this bird have only been found in Cuba.

Cuba is a collection of tightly packed islands in the Caribbean Sea. As we have seen, islands are treasure troves of biological diversity as any animal that somehow manages to reach an isolated island can evolve independently of its relatives on the mainland. Long ago, Cuba was home to a unique collection of animals that evolved from North American and South American immigrants. One of the most bizarre Cuban animals was the giant owl.

The remains of this bird were first discovered in Cueva de Pio Domingo in western Cuba, and it was thought, initially, that they belonged to a Cuban species of terror bird because of their size. The bones clearly belonged to a large bird that spent most of its time on the ground. In the early 1960s, a paleontologist was examining these bones, and he saw them for what they really were: the remains of a giant, extinct owl. Today around 220 owls species are recognized and zoologists separate them into two groups: the typical owls and the relatively long-legged and highly nocturnal barn owls. For most owls, the day begins when the sun goes down, when they leave their daytime retreats to hunt their prey. There can be few predators as beautifully adapted as the owls. Their senses of sight and hearing are acute, and their wing beat is muffled by the soft barbule tips on the leading edge of the flight feathers, which dampen air noise during flight. These adaptations allow them to find prey in low light levels and to make an approach without alerting the hapless victim.

Cuban Giant Owl—The ground-dwelling Cuban giant owl stood about 1 m high, dwarfing most modern owls. (Renata Cunha)

Since the first bones of the giant owl came to light, lots of remains have been found all over Cuba, including three more or less complete skeletons. These bones indicate a large animal that was predominantly a ground dweller. Isolated on the island of Cuba, the giant owl deviated from the owl norm and took up life on the ground. Although there are owls today that spend a lot of time on the ground (e.g., the burrowing owl, *Athene cunicularia*), they still have large wings and powerful flight muscles and can take to the air with ease. Unlike some other ground-dwelling birds that have completely forsaken the power of flight, the sternum of the giant owl does have a keel, indicating that the living bird's flight muscles may have been large enough to take the bird into the air for very short distances. Much like a turkey, the giant owl was probably capable of short, feeble flights when threatened, but its long legs and large feet suggest that it preferred stalking around at ground level. In terms of size, the giant owl was far in excess of any living owl. The two eagle owl species, (*Bubo bubo* and *Bubo blakistoni*), are the largest living owls and can reach a weight of around 4.5 kg. The Cuban owl was probably double this weight. Because of its size and because Cuba was free of large mammalian predators, the Cuban giant owl may have switched from a nocturnal lifestyle to a diurnal one. Strutting around the forested islands of Cuba, the giant owl used its predatory adaptations to hunt animals as large as hutia (*Capromys pilorides*), stocky Caribbean rodents and small capybara (*Hydrochaeris hydrochaeris*), the largest living rodents. Like other ground-dwelling birds, the Cuban giant was probably an accomplished runner, and it very likely ran its quarry down before dispatching it with its powerful talons and beak.

The owls we know today usually build their nests in lofty places that afford the eggs and young some protection from predators. Tree holes and other cavities are favored nesting sites, but these must have been out of the question for the giant owl. Even if it could have reached a tree hole, there were probably few of a sufficient size to accommodate its large body. The only option was a nest on the ground or in a burrow, and fortunately, there were few, if any, Cuban animals to prey on the eggs and young of the giant owl. The presence of two giant birds guarding the nest must have been more than enough to discourage even a hungry opportunist.

Exactly when the ancestors of the giant owl colonized Cuba is a mystery, but the descendents of these nocturnal hunters evolved over hundreds of thousands of years to fill the niche of a large, diurnal, ground-dwelling predator. The prehistoric Cuba must have been a paradise, but once again, humans arrived, bringing with them devastation and extinction. The first humans to reach Cuba arrived from South America, Central America, and North America in a complex series of migrations as long as 8,000 years ago. These people, known to anthropologists as the Taíno and Ciboney, took up residence and practiced hunter-gathering and agriculture. Ground-dwelling birds that have evolved on isolated islands have absolutely no defense against humans. The giant owl, at around 9 kg, was a considerable source of animal protein and one that was easy to catch. Although the islands of Cuba have quite a large land area, the giant owl, as top predator, could never have existed in huge numbers. Human hunting as well as habitat destruction must have decimated the populations of this bird, and the animals the humans brought with them made short work of the eggs and nestlings of this amazing owl. The youngest remains of the giant owl are around 8,000 years old, and it is very unlikely that a large, cursorial bird could have persisted for anything more than a couple of centuries after humans reached Cuba.

+ In appearance, the giant owl is thought to have resembled a large burrowing owl (*Athene cunicularia*), but its remains show that it was actually more closely related to the wood owl (*Strix* sp.).
+ A very unusual and extremely rare animal called the "Cuban solenodon" (*Solenodon cubanus*) still manages to cling to survival on these Caribbean islands, but since its discovery in 1861, only 37 specimens have been caught. This odd, nocturnal, burrowing creature, one of the few venomous mammals, is a reminder of the days when Cuba was populated by odd animals, creatures which evolved in isolation on these tropical Caribbean islands (see the entry for Marcano's solenodon in chapter 3).

Further Reading: Brodkorb, P. "Recently Described Birds and Mammals from Cuban Caves." *Journal of Paleontology* 35 (1961): 633–35.

Extinction Insight: Entombed in Tar—The Rancho La Brea Asphalt Deposits

In downtown Los Angeles is one of the most fantastic fossil sites in the whole world—a place that has given us an unparalleled glimpse of a small corner of ice age earth. Rancho La Brea, frequently referred to as the La Brea tar pits, has yielded around 1 million bones since excavations began there in 1908. The site is actually above an oil field, and oil has been seeping to the surface through fractured rocks for 38,000 years. When the oil reaches the surface, the more volatile chemicals evaporate, leaving a heavy, thick tar (asphalt).

For millennia, Amerindians used the tar from the asphalt pools for waterproofing shelters and canoes as well as for glue. It was even considered valuable enough to be traded. A Franciscan friar, Juan Crespi, makes the first written mention of the asphalt deposits during his expedition with Gaspar de Portola (the first Spanish governor of the Californias) in 1769–1770. Later, the site was part of an 1,800-hectare Mexican land grant given to Antonio Jose Rocha in 1828. It then found its way into the hands of the Hancock family, and Captain George Allen Hancock donated the 23 acres of Hancock Park to Los Angeles County in 1924.

The oil beneath the asphalt deposits is itself a fossil, the oily, organic remnants of the tiny organisms that make up marine plankton. Between 5 and 25 million years ago, this part of California was actually a shallow sea, and these single-celled organisms died and sank to the bottom, where they became part of a thick layer of sediment. As the climate changed and the continents moved around, the sea disappeared, and the dead plankton, entombed under tonnes of overlying sediment, were slowly converted into an oil and gas deposit by the pressure and heat. Fractured rocks above this oil field provided a path to the surface, through which the crude oil seeped, accumulating in numerous pools that dotted what is today known as Hancock Park. Over thousands of years, some of these seeps ended, while new ones began, but all the while, they were a trap for a myriad of species of animals. Sometimes a seep produced a tar pit that was deep enough to trap really large animals. How animals became trapped in the tar isn't known for sure, but it is thought that water, leaves, and dust accumulated on the tar pits and animals were deceived into wading in to bathe or drink. This was the last mistake the animal made, as the sticky tar snared its legs and made escape impossible. The commotion caused by struggling animals and the smell of dead animals that had already perished in the sticky goo attracted the attention of predators and scavengers. Not only are the number and diversity of the fossils from La Brea unprecedented, but it is the only fossil assemblage on earth where predators outnumber prey. This is because a large animal, like a mastodon, struggling in the tar attracted numerous predators and scavengers, all of which were keen to get their teeth and claws into the doomed beast, and in trying to do so, some of them also met their end in the tar. It is even possible that prey and predators became trapped during a chase that ended badly for all parties concerned. This might seem unlikely, but a major entrapment like this only needed to happen once every 10 years over a 30,000-year period to account for all the bones in the asphalt deposits. The dead bodies would sink into the tar, and as the seep stopped, the volatile elements of the oil continued to evaporate, leaving hard, asphalt-impregnated clay and sand, and the bones.

Even before the paleontological importance of this site was recognized, ranchers took notice of the bones protruding from the asphalt deposits but mistakenly believed them to be the remains of cattle and pronghorn that had wandered into the sticky tar. To date, more than 660 species of plant and animal have been found in the asphalt deposits, all of which got trapped in the tar between 8,000 and 38,000 years ago.

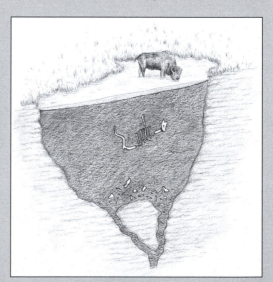

Rancho La Brea Asphalt Deposits—This is an example of how an animal met its end in the asphalt deposits. The bison is attracted to the seep to drink the water that has pooled on the asphalt. It becomes trapped in the sticky asphalt and dies. Its bones sink into the asphalt, and after thousands of years, humans find and excavate them. (Phil Miller)

The discoveries include 59 species of mammal and 135 species of bird. Some of the animals recovered from the pits are extinct—for example, the saber tooth cat *Smilodon fatalis*, ground sloths, mastodons, and mammoths—while others are still around today, for example, pronghorn (*Antilocapra americana*), elk (*Cervus Canadensis*), coyote (*Canis latrans*), and bobcat (*Lynx rufus*). The asphalt deposits have yielded the remains of more than 1,600 dire wolves and around 1,200 saber tooth cats. Apart from being stained brown after several millennia entombed in tar, the asphalt deposit bones are brilliantly preserved, and in some cases, tar has seeped into the cranial cavity of the skull to produce an endocast of the brain.

The bones of the large extinct animals discovered in the asphalt deposits have always attracted the most attention, especially the perfectly preserved skulls of the saber tooth cat, but relatively recently, scientists realized that alongside these bones were fossil seeds, pollen, insects and mollusks, and the bones of fish, amphibians, small birds, and rodents. These microfossils allow paleontologists to build up a very detailed picture of the habitat and climate in Los Angeles during the final part of the last ice age.

Interestingly, of all the bones recovered from Rancho La Brea, only one human skeleton has been found: a 1.5-m-tall woman in her mid-twenties, who appears to have suffered a blow to the head.

5

10,000–12,500 YEARS AGO

GLYPTODONT

Glyptodont—The glyptodont's huge, domed carapace made it almost invulnerable to predators. (Natural History Museum at Tring)

Glyptodont—Even if a predator was stupid enough to attack an adult glyptodont, its lashing tail could inflict some serious injury. (Natural History Museum at Tring)

Scientific name: *Glyptodonts*
Scientific classification:
    Phylum: Chordata
    Class: Mammalia
    Order: Cingulata
    Family: Glyptodontidae
When did it become extinct? The glyptodonts became extinct about 10,000 years ago.
Where did it live? The glyptodonts were native to South America, although fossils of a similar animal are known from the southern parts of North America.

South America was once home to a number of glyptodont species, all of which looked like enormous armadillos. These are surely among the most bizarre animals that have become extinct in the last few thousand years, and some of them reached huge sizes. An adult *Glyptodon*, the typical representative of this group, which used to amble around in Argentina, could have been 4 to 5 m in length and weighed in excess of 2,000 kg, making it as big as a small car.

The short, squat limbs and fused vertebrae of the glyptodonts supported a massive, domelike carapace that must have afforded the living animal a formidable level of protection from hopeful predators. This carapace was composed of more than 1,000 bony plates, each of which was more than 2 cm thick. The head was also heavily protected with a bony plate, as it could not be withdrawn into the carapace like that of a turtle. Not only were the glyptodonts heavily armored, but they also had a fearsome weapon in the shape of their tail. In some species, this was fortified with rings of bony plates, whereas other species sported a thuggish club or dangerous looking, macelike growth. Any predator would have to have been wary of a glyptodont's lashing tail if it were to survive to see another day.

The level of protection displayed by the glyptodont came at a price because it was very heavy indeed. The short, squat legs would only have been able to propel the great bulk of the beast at a very lumbering pace. Bones and the way they fit together allow scientists to estimate how the living animal moved. Simulations of a glyptodont's gait show that it would be struggling to amble along at anything more than around 4 to 5 km per hour. Bones can also give us insight into how the animal went about its everyday life. Glyptodonts only had teeth in the rear of their mouths, but they continued to grow throughout the animal's life. This and the massive, deep mandible, which, in the living animal, was moved by huge jaw muscles, show that the glypotodonts were herbivorous animals that fed on fibrous plant food. Exactly what plants they ate can only be surmised, but perhaps the grasses and low-growing vegetation of the prehistoric South American grasslands were their favored food.

It is not clear what predators the glyptodont's armor was protecting them from. Certainly the fossil record has not offered up any predator that appears to have been powerful enough to kill an adult glyptodont. Saber tooth cats, huge terror birds, and jaguar-sized predatory marsupials all lived alongside the glyptodonts, but it is hard to believe that any of these animals could have gotten the better of an adult glyptodont. Perhaps only the young glyptodonts were vulnerable to predation, as is the case for some of the large mammals that wander the savannah of Africa today.

The causes for the demise of the glyptodonts can only be guessed, but it is extremely likely that they succumbed to habitat changes brought about by shifts in the earth's climate. These bizarre animals appear to have disappeared around 10,000 years ago—around the time the last ice age was coming to an end. It is likely that the first humans in South America hunted glyptodonts, but it is doubtful that this was the cause of their extinction. The glypotodonts' numbers probably declined in the face of changing habitats and it is possible that human exploitation hastened their demise.

+ The glyptodonts first appear in the fossil record in the Miocene, which spanned a period of time from 5 to 23 million years ago. They are thought to have evolved from an armadillo-like animal, subsequently diversifying and reaching large sizes.

+ The glyptodonts shared South America with a huge number of very large animals, all of which are collectively known as megafauna. All of the really large representatives of the megafauna became extinct around 10,000 years ago, which is further evidence that there were some global changes occurring, although hunting by prehistoric humans can never be ruled out.

+ The formidable carapace of the glyptodonts and the intriguing tail weaponry of some species may have been put to good use in fights between males during the breeding season.

+ North America, like South America, had its own megafauna, but the two groups of animals on these huge landmasses were isolated from one another until a great deal of geological activity formed the isthmus of Panama, effectively joining the two continents around 3 million years ago. This land bridge allowed animals to move between the landmasses, an event known as the Great American Interchange (see the "Extinction Insight" in chapter 2). The glyptodonts took advantage of this bridge and crossed into North America, eventually spawning the species known as *Glyptotherium texanum*, whose fossils are found throughout Texas, South Carolina, and Florida.

+ It is highly likely that the first humans to reach the Americas saw the glyptodonts alive, but we don't know the extent to which hunting affected their numbers. What we know for sure is that certain tribes from Argentina were intimately aware of the animal's fossils. It is said that certain tribes used the huge carapaces as shelters during bad weather. Indeed, the animal still exists in the folk memory of some of these peoples.

+ Although fossils can tell us a lot about what an animal looked like and how it lived, the bare bones often only give us tantalizing glimpses of the living animal. One such mystery is the glyptodont's reduced nasal passages, which appear to have served as anchoring sites for considerable muscles. This observation has led some people to suggest that the glyptodonts were equipped with some manner of trunk, but as with many paleontological mysteries, we will never know for sure.

Further Reading: Haines, T., and P. Chambers. *The Complete Guide to Prehistoric Life*. Richmond Hill, ON Canada: Firefly Books, 2006; McNeill Alexander, R., R. A. Farin, and S. F. Vizcaíno. "Tail Blow Energy and Carapace Fractures in a Large Glyptodont (Mammalia, Xenarthra)." *Zoological Journal of the Linnaean Society* 126 (1999): 41–49.

SABER TOOTH CAT

Scientific name: *Smilodon populator*
Scientific classification:
   Phylum: Chordata
   Class: Mammalia
   Order: Carnivora
   Family: Felidae
When did it become extinct? This cat is thought to have gone extinct around 10,000 years ago, but as with any prehistoric animal, it is impossible to know exactly when it disappeared.
Where did it live? This feline lived in South America.

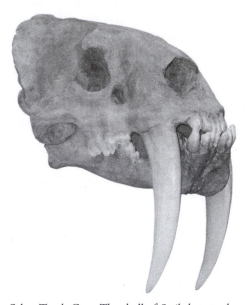

Saber Tooth Cat—The South American *Smilodon populator* was the largest saber tooth cat as well as one of the largest cats that has ever lived. (Renata Cunha)

Saber Tooth Cat—The skull of *Smilodon populator* clearly shows the enormous canines of this formidable extinct cat. It used these teeth to inflict fatal wounds on some of the large South American herbivorous mammals. (Ross Piper)

The *Smilodon* species, often called saber tooth cats, are among the most famous of all prehistoric beasts, and the species described here was the biggest and most powerful of them all. The Latin name of this cat, *Smilodon populator*, can be translated as the "knife tooth that destroys." Fully grown, *S. populator* was the same height and length as a large lion, but much heavier. They were around 1.2 m at the shoulder and may have reached 400 kg—heavier than any big cat alive today. Unlike modern big cats, *S. populator* had a very stubby tail, and it also had very robust and heavily muscled forequarters—an important adaptation for catching and subduing prey. The bones of *S. populator*'s forelimbs were relatively short and quite broad, indicating that they were attached to some very powerful muscles. These worked together with the muscles in the shoulders and back to provide tremendous force. Without doubt, the most impressive feature of *S. populator* is the massive canine teeth in the upper jaw. They were huge—far bigger than any tooth that has graced the mouth of any cat before or since. These formidable curved fangs were around 20 cm long, and to accommodate them, the mouth could open extraordinarily wide, up to 120 degrees (a modern lion's maximum gape is about 65 degrees).

Why did *S. populator* have such monstrous canines? We know that this predator stalked the earth at the same time as many species of large herbivorous mammals, but it is very unlikely that *S. populator* was capable of subduing the adults of the Pleistocene giants: mammoths, mastodons, giant sloths, and the like. However, the young of these giants and a host of other herbivores were well within the predatory abilities of the saber tooth cats, and they represented a feast for any animal that could bring them down. Catching and killing a large herbivore is no mean feat, even for a hugely powerful, 400-kg cat with 20-cm canines.

Exactly how *S. populator* and the other *Smilodon* species caught and killed their prey has been a bone of contention for decades, but a look at the remains of these long-dead animals does give us some clues. Their stocky build and their relatively short limbs indicate that they were probably ambush predators. They may have skulked behind bushes and other vegetation and pounced on an unfortunate ungulate when it came within range. This is a plausible explanation of how they caught their prey, but how did they kill? For some time, it was thought that these cats used their canines to prize apart the prey's vertebrae, but research has shown that their teeth were much too brittle for this. If the jaws were slammed shut on bone, the canines would have shattered, and without its weapons, a saber tooth would have starved to death. It was also suggested that the teeth were used to slice open the soft underbelly of the prey, but again, the risk of contacting bone during the killer bite was too great. It seems that the *Smilodon* species actually went for the neck. Using the great muscular strength in their forelimbs to keep hold of the victim long enough to deliver the killer bite, they plunged their huge fangs into the soft throat of the prey, severing the important blood vessels and crushing the windpipe. Biting this way, a large fold of the prey's skin was probably taken in to the cat's mouth, some of which may have been torn away as the feline pulled away. In this scenario, the prey died quickly from blood loss and suffocation, and the cat could have dug in to its meal quickly. It is very likely that *S. populator* fed on the same sized animals that lions and tigers are capable of dispatching today—it's just that it killed in a different way.

As with many of the amazing mammals that became extinct at around the end of the last glaciation, we can never be certain of what led to the demise of these cats. We do know that the habitats in which these animals evolved went through massive changes as the climate went through cyclical periods of cold and warm, but this alone is not enough to explain the disappearance of these felines. It is interesting to note that the spread of humans around the world appears to coincide with the disappearance of these intriguing cats and many other prehistoric, predatory mammals. Perhaps a combination of climate change and hunting by prehistoric humans pushed the populations of the large herbivores to extinction. As their prey dwindled, the *Smilodon* species, with their very specialized hunting technique, found it increasingly difficult to find sufficient food in the changing landscape. It is amazing to think that our ancestors probably watched the *Smilodon* species hunting and going about their everyday lives. Even more intriguing is the possibility that our forebears were probably killed and eaten by these impressive cats.

- Three species of *Smilodon* are known: *S. populator*, *S. fatalis*, and *S. gracilis*. The species described here, *S. populator*, probably evolved from *S. gracilis* after it reached South America from the north. *S. gracilis* probably also gave rise to *S. fatalis*, which is the most well known of these animals as bones from at least 1,200 individuals have been found in the asphalt deposits of Rancho La Brea in Los Angeles (see the "Extinction Insight" in chapter 4).
- The asphalt-stained and well-preserved bones of the Rancho La Brea pits tell us a great deal about these animals, including the afflictions and diseases that troubled them. The *S. fatalis* bones show evidence of infections, healed breakages, muscle damage, osteoarthritis, and even wounds inflicted by others of their kind. Torn muscles and ligaments show that these cats used enormous force when attacking their prey. These remains give us an unparalleled glimpse of the earth many thousands of years ago (see the "Extinction Insight" in chapter 4).
- The *Smilodon* species and the other saber tooth cats are sometimes mistakenly called "saber tooth tigers." They are cats, no question, but they are not closely related to the tigers we know today.

+ Some *Smilodon* bones have been found in situations that have led some scientists to suggest that they were social and hunted in groups. The brains of *S. fatalis* are very similar in size and structure to similarly sized modern-day cats, and we are very familiar with the complex social behavior of the lions. There is no direct evidence for well-developed social behavior in any *Smilodon* species, but it is possible that they lived in groups and depended on teamwork to catch their prey.

+ It was once thought that the *Smilodon* species had quite a weak bite, but recent research suggests that their bite was probably as powerful as that of the largest modern big cats. They could also probably use their neck muscles to drive their teeth through the tough hides of their prey.

Further Reading: Barnett, R., I. Barnes, M.J. Phillips, L.D. Martin, R. Harington, J.A. Leonard, and A. Cooper. "Evolution of the Extinct Sabertooths and the American Cheetah-like Cat." *Current Biology* 15 (2005): 589–90; Christiansen, P., and J.M. Harris. "Body Size of *Smilodon* (Mammalia: Felidae)." *Journal of Morphology* 266 (2005): 369–84; McCall, S., V. Naples, and L. Martin. "Assessing Behavior in Extinct Animals: Was *Smilodon* Social?" *Brain, Behaviour and Evolution* 61 (2003): 159–64; Christiansen, P. "Comparative Bite Forces and Canine Bending Strength in Feline and Sabertooth Felids: Implications for Predatory Ecology." *Zoological Journal of the Linnean Society* 151 (2007): 423–37; Anyonge, W. "Microwear on Canines and Killing Behavior in Large Carnivores: Saber Function in *Smilodon fatalis.*" *Journal of Mammalogy* 77 (1996): 1059–67.

SCIMITAR CAT

Scimitar Cat—The scimitar cats are another extinct species of felines with large canine teeth. They were large, long-limbed animals, and they probably used their impressive teeth to kill and dismember large herbivorous mammals. (Renata Cunha)

Scientific name: *Homotherium* sp.
Scientific classification:
    Phylum: Chordata
    Class: Mammalia
    Order: Carnivora
    Family: Felidae

When did it become extinct? The estimates for when the last scimitar cat became extinct vary between 10,000 and 13,000 years ago, but it is possible that they survived into more recent times.

Where did it live? The remains of the scimitar cat have been found in North America, Eurasia, and Africa.

Thousands of years ago, the world was a dangerous place, what with the saber tooth cats on the prowl—predators that must have surely been greatly feared by our ancestors. If fearsome saber tooths were not enough, there were other species of powerful, large-fanged cats that stalked the earth at the same time, and among the most well known of these are the scimitar cats.

Fossils of scimitar cats are not as common as those of the saber tooth cats, but the remains of 33 adults and kittens of one species (*Homotherium serum*) were found in a cave in Texas. Some of these skeletons were complete, giving us a good idea of what the scimitar cats looked like as well as throwing some light on how they lived.

The scimitar cats were around the same size as a modern lion, with a stumpy tail; however, they were lightly built, with relatively long limbs. Like the spotted hyena (*Crocuta crocuta*), their forelegs were noticeably longer than their hind legs, and as a result, their backs sloped toward the rear. Although their forelegs were quite slender compared to the *Smilodon* saber tooth cats, they were undoubtedly powerful and used to great effect when grappling with prey. As well as long limbs, the scimitar cat's claws could be retracted as much as those of a modern-day tiger or lion. The ability to retract their claws has important implications for the way these cats caught their prey, which will be covered in more depth later. The serrated canines of the scimitar cats were not as large as the massive daggers of the saber tooth cats, but they were still impressive weapons. The European species, *H. crenatidens*, has the biggest canines of all the known scimitar cats. At around 100 mm, they dwarf those of an adult tiger, which are normally 55 to 60 mm long. To protect these fangs, the mandible of the scimitar cat was massively developed, with flanges that acted like scabbards, probably to protect the canines. These scabbards were at their most impressive in *H. crenatidens*. Not only were the canines of scimitar cats fearsome, but their incisors were equally arresting. In *H. crenatidens*, the incisors undoubtedly formed an effective puncturing and grabbing mechanism that tore a lump of flesh from the unfortunate victim and were useful for carrying dismembered limbs.

What can the remains of the scimitar cats tell us about the way they lived? We know from where scimitar cat bones have been found that these predators probably migrated with the cyclical periods of cold and warm that have prevailed on earth for hundreds of thousands of years, and it is likely that they roamed the cold expanses and forests of the Northern Hemisphere. We can assume that the conditions in which these animals survived are very similar to what we see in the Northern Hemisphere today. Much like the Siberian tiger, the scimitar cats were adapted to cold, temperate conditions. For camouflage, they may have had very pale, dappled fur, much like a lynx (*Felis lynx*) or bobcat (*Lynx rufus*). A large predator with dark fur in this environment would have stood out like a beacon, thus making it very difficult to approach wary prey.

The long legs and the large nasal cavity of these cats have led some scientists to suggest that they could pursue their prey over long distances. These felines were undoubtedly capable of short sprints, but it's very unlikely that they were capable of long-distance pursuits. Like almost all other cats, the scimitar cats were probably ambush predators, using stealth to get within striking distance before launching a lightning attack. Interestingly, the structure of the scimitar cat's rear suggests that they were not very good at leaping. The large nasal cavity probably also served to warm incoming air before it went into the lungs.

As the teeth of the scimitar cats are very different from those of the saber tooth cats, it has been argued that the former had a distinct killing technique to that used by its bulky relative. With this said, we can never be sure how the scimitar cats caught their prey, but the amazing haul of bones discovered in Friesenhahn Cave, Texas, includes a huge number of bones from what could have been prey animals. This unprecedented haul includes lots of milk teeth from more than 70 young mammoths. Could the scimitar cat have been a specialist predator of young mammoths? Based on observations of elephants, we know that youngsters aged between two and four years old will stray from the family group to satisfy their curiosity with the world around them. Isolated, they are vulnerable to attack from lions. It is possible that the scimitar cat was preying on similarly curious young mammoths and maybe even dismembering the carcasses before certain parts were taken back to the cave for consumption by the adults and cubs. These mammoth remains may have been brought into the cave by other animals, such as dire wolves, the remains of which have also been found in this refuge; nevertheless, we are left with a tantalizing glimpse of how these long-dead cats may have lived. Perhaps they were specialist hunters of the young of the numerous elephantlike animals that once roamed the Northern Hemisphere.

+ The scimitar cats lived throughout Europe, North Africa, and Asia. There is also some fossil evidence that they reached South America.

+ The canine teeth of the scimitar cats appear to be adapted for slashing flesh, rather than for stabbing, which was the tactic of the saber tooth cat. When the scimitar cat's mouth was closed around the throat of an unfortunate victim, the canines formed an effective trap along with the incisors. As the cat pulled back from the prey, it probably ripped out a sizeable chunk of skin, fat, and muscle, causing rapid blood loss.

+ Apart from the Friesenhahn Cave bones (discovered during the summers of 1949 and 1951; see the "Extinction Insight" in chapter 1), remains of the scimitar cat are relatively rare, and other finds are generally of a disjointed bone or two. The rarity of specimens suggests that the scimitar cats may have been quite uncommon, albeit widespread, predators that stalked the Northern Hemisphere up until the end of the ice age.

+ As with the last saber tooth cats, we cannot be certain what caused the demise of the scimitar cats, but we cannot rule out the effect of humans hunting the prey of these animals, eventually depriving them of food.

+ The Pleistocene abounded with a variety of big cats, but today, there are only eight species of big feline. The Americas have lost all of their big cats, except the cougar (*Puma concolor*) and jaguar (*Panthera onca*).

Further Reading: Reumer, J.W.F., L. Rook, K. Van Der Borg, K. Post, D. Mol, and J. De Vos. "Late Pleistocene Survival of the Saber-Toothed Cat *Homotherium* in Northwestern Europe." *Journal of Vertebrate Paleontology* 23 (2003): 260–62; Mauricio Antón, M., A. Galobart, and A. Turner. "Coexistence of Scimitar-Toothed Cats, Lions and Homininss in the European Pleistocene: Implications of the Post-cranial Anatomy of *Homotherium latidens* (Owen) for Comparative Palaeoecology."

Quaternary Science Reviews 24 (2005): 1287–1301; Antón, M., and A. Galobart. "Neck Function and Predatory Behavior in the Scimitar Toothed Cat *Homotherium latidens.*" *Journal of Vertebrate Paleontology* 19 (1999): 771–84.

AMERICAN MASTODON

American Mastodon—The American mastodon was an elephantlike creature that inhabited North America for a much longer period of time than the mammoths. (Renata Cunha)

Scientific name: *Mammut americanum*
Scientific classification:
    Phylum: Chordata
    Class: Mammalia
    Order: Proboscidea
    Family: Mammutidae
When did it become extinct? The American mastodon is thought to have become extinct around 10,000 years ago.
Where did it live? The American mastodon was native to North America, and many remains have been found in the area immediately south of the Great Lakes.

Thousands of years ago, several species of mammoth could be found on the North American continent; however, these were not the only huge, shaggy, elephantlike beasts to be found in these lands. The mastodon, a creature that is often confused with the mammoth, lived in North America for a very long period of time—much longer than the mammoth—evolving from creatures that crossed into the New World from Asia via the Bering land bridge as early as 15 million years ago.

This enigmatic, long-dead mammal looked very much like its distant relative, the mammoth, but it was not as large as the largest of these animals, reaching a height of around 3 m, a length of about 4.5 m, and a weight of 5.5 tonnes. Its skeleton was stockier, with shorter, more robust legs than a similarly sized mammoth, and its skull was also a different shape, giving the mastodon a receding brow, rather than the big, flat forehead of their elephantine relatives. The tusks of the mastodon were very impressive, reaching lengths of around 5 m, but they were not as curved as the mammoth's. Like the mammoths, the mastodons were covered in thick, shaggy fur that was needed to ward off the cold, but it is impossible to know what color this pelage was in life—dark brown has been suggested, but we have no way of knowing. So, on the outside, the mastodons and the mammoths were very familiar, and the best way to tell them apart is to look at their teeth. The teeth of a mammoth are topped off with shallow enamel ridges, making them very effective grinding surfaces for the mashing up of grasses and other coarse plant matter. The mastodon's teeth, on the other hand, are quite different, as each one is surmounted with a small, enamel-covered cone that looks a lot like a nipple, which is where the Greek name *mastodon* comes from (*mastos* translates as "breast"; *odont* translates as "tooth").

The structure of the mastodon's teeth gives us an idea of what these animals ate. As the teeth lacked a ridged grinding surface, we can assume that plants like grasses were off the menu for these lumbering beasts, but their dentition seems to be well suited to chopping and chewing twigs and leaves. Unlike the mammoths, which were grazing animals, the mastodon must have been a browser, feeding in the same way as modern elephants can sometimes be seen doing in the African bush—pulling branches to their mouth with their prehensile trunk. The fact that these animals fed in a different way to the mammoths is the reason why they were able to live alongside one another on the same landmass for thousands of years without coming into competition. As many mastodon remains have been found in lake deposits and in what were once bogs, it has been suggested that they spent a lot of their time in water, wading through the shallows grasping at succulent foliage with their flexible trunks.

The sharp eyes of an expert can reveal lots of telltale signs that enable us to build a picture of how the animal lived, and the remains of the mastodon are no exception. The tusks of male mastodons have been shown to bear interesting pits on their lower sides that occur at regular intervals. It has been proposed that these marks are scars, evidence of the damage caused by fights between males during the breeding season. Male mastodons must have locked tusks with the intention of driving the tip of their weapons into the heads or flanks of their opponents, incapacitating or even killing them. These violent struggles forced the underside of the tusk against its socket, damaging a point on the adornment that was revealed as it grew. Annual fighting led to a series of scars on the tusk. This is only a theory, but it offers a tantalizing insight into the behavior of these long-extinct giants.

What became of the mastodon? How come North America is no longer home to these great beasts? The honest answer is that we simply don't know; however, numerous theories attempt to explain their disappearance. Climate change has been cited as a culprit, even though the mastodons survived for millions of years through numerous cycles of global cooling and warming. A second theory is that humans hunted the mastodons to extinction during their dispersal into North America from eastern Asia 15,000 to 20,000 years ago, at the end of the latest ice age. We know that humans hunted these animals as their weapons have been found with mastodon remains. A mastodon skeleton has even been found with a spear point embedded in the bone, and even more remarkably, the individual in question managed to survive the attack as the wound had healed. Such finds tell us that our forebears hunted these animals, but they give us no idea of the intensity of this predation. A third theory is that tuberculosis drove the mastodons over the edge. Again, the tale of the bones shows that mastodons did indeed suffer from this disease, but was it enough to drive them to extinction? A plausible explanation for their disappearance is a combination of all these factors. Climate change may have put a lot of pressure on the population of these animals, and disease may have weakened them still further, with hunting bringing the final death knell.

- The ancestors of the mastodons evolved in North Africa around 30 to 35 million years ago. From this point of origin, they spread through Europe and Asia, eventually crossing into North America.
- Europe was once home to a species of mastodon, but it became extinct around 3 million years ago, leaving North America as the last refuge for these animals.
- As the remains of mastodons are found singly, it has been proposed that these animals did not form family groups. They may have led a solitary existence, only coming together during the breeding season, which is in contrast to modern elephants, and probably mammoths.
- The disease tuberculosis leaves characteristic grooves on the bones of infected animals. It is possible that diseases such as tuberculosis were brought to North America by humans as they dispersed throughout the continent.

Further Reading: Fisher, D.C. "Mastodon Butchery by North American Paleo-Indians." *Nature* 308 (1984): 271–72; Dreimanis, A. "Extinction of Mastodons in Eastern North America: Testing a New Climatic-Environmental Hypothesis." *Ohio Journal of Science* 68 (1968): 337–52.

GIANT BEAVER

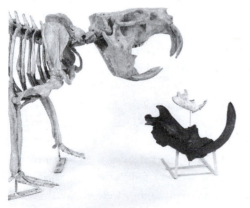

Giant Beaver—This picture shows the giant beaver's skull and mandible compared to that of a modern American beaver. The difference in size is startling. (Richard Harrington)

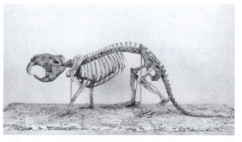

Giant Beaver—The giant beaver was about the same size as a modern black bear. (Richard Harrington)

Scientific name: *Castoroides ohioensis*
Scientific classification:
　　Phylum: Chordata
　　Class: Mammalia
　　Order: Rodentia
　　Family: Castoridae

When did it become extinct? The giant beaver is thought to have become extinct around 10,000 years ago.

Where did it live? This giant rodent lived in North America. Its remains have been found from Florida to the Yukon and from New York State to Nebraska.

Modern-day beavers are big by rodent standards, with a weight of up to 35 kg for the European species (*Castor fiber*). Imagine, then, a prehistoric beaver that weighed around 200 kg and was around 2.5 m long—about the same size as a black bear (*Ursus americanus*). This was the giant beaver, and it was one of the largest rodents that has ever lived. Unlike extinct beasts, such as the mammoth and cave bear, the giant beaver has never been found depicted in cave paintings, so we can only make assumptions of its appearance in life based on its bones. In general appearance, the giant beaver was very similar to the modern-day species, just a lot bigger. Like the living beaver, this giant had its eyes high on its head so that it could see above the water when the rest of its body was submerged. The front incisors of the giant beaver were massive (about 15 cm long), relatively much larger than the incisors used by the living American beaver to gnaw through young trees. Unlike modern beavers, the front edge of the giant beaver's incisors was not smooth; instead, it was heavily ridged, and it has been proposed that these structures strengthened the very long teeth, protecting them from breakage when they were being used.

How did the giant beaver use these impressive teeth? Some experts believe that the teeth were for gnawing at wood, while others think that gouging was more likely. The giant beaver must have done some tree gnawing because if its modern-day relatives are anything to go by, nibbling wood is one way of keeping the ever-growing incisors in check. Like its surviving relatives, the giant beaver probably got some of its sustenance from eating bark to supplement the nutrients it obtained from eating aquatic vegetation.

The modern-day beavers love water and spend a lot of their time in lakes and rivers, but they are also very mobile on land and often travel good distances on foot from one lake to another. It has been suggested that due to its great size, the giant beaver may have been slow and clumsy on land; therefore it may have been predominantly an aquatic animal, only leaving the water to search for food. With that said, the immense bulk of the hippopotamus (*Hippopotamus amphibius*) does not stop it leaving the water to graze at night.

Where exactly did the giant beaver live? Many giant beaver bones have been found in old swamp deposits, so we can assume that this giant rodent preferred lakes surrounded by swamp, and it seems to have flourished in an area around the Great Lakes. Three almost complete skeletons have been found in Indiana and Minnesota. Toward the end of the last ice age, this region was dotted with numerous swamps and lakes—the probable preferred habitat of this giant rodent. The density of lakes, marshlands, rivers, and streams probably lent itself to the dispersal of an animal that was not fond of leaving the water. Giant beaver remains have been found over a very large area, so they were obviously occupying an ecosystem rich in aquatic habitats. Even if the giant beavers rarely moved far over land, they could have dispersed over great distances by traveling between the extensive network of interconnected lakes that once studded North America, the remnants of which we still see today.

The living beavers are unique for their building abilities. They construct lodges of saplings, branches, and twigs to live in and dams that curb the flow of rivers and streams.

These industrious efforts can change whole habitats. Did the giant beaver do the same, constructing enormous structures of saplings and cut wood? We have no way of knowing for sure, but in 1912, part of a young giant beaver's skull and its possible lodge were discovered near New Knoxville in Ohio. The lodge was said to have been 1.2 m high and 2.4 m across and was built from saplings with a diameter of 7.5 cm.

Like many of the other great beasts that once roamed North America, the giant beaver became extinct around 10,000 years ago. The exact cause of its demise is a mystery. As a species, the giant beaver survived for around 2 million years, and in that time, glaciers expanded and retracted as the earth's climate oscillated between longer cold and shorter warm periods for at least 10 cycles. The giant beaver survived all of these oscillations and the changes they brought, except the last one. Humans have been implicated in the extinction of the North American megafauna as there is thought to be a link between the spread of the prehistoric human population and the disappearance of the American continent's giant beasts, but there is no direct evidence that humans hunted the giant beaver. With that said, a 200-kg animal with lots of meat on its bones and a dense pelt that could have been made into warm clothing must have been coveted by prehistoric North Americans.

- The first remains of this animal were found near Nashport, Ohio, in a peat bog, and they were described as belonging to a giant beaver in 1838.
- The giant beaver lived alongside the modern-day American beaver (*Castor canadensis*). For two similar species to coexist, there must have been differences in the habitats they preferred or possibly in the food on which they depended. Perhaps the giant beaver, with its capacious mouth, was able to use larger trees for food and building, while its small relative could nibble away at smaller saplings.
- Even though North America still has its fair share of wilderness, it's hard to imagine what it must have looked like thousands of years ago, long before the advent of intensive agricultural and urban development. For millions of years, it was one vast wilderness completely untouched by humans, where the forests, plains, lakes, rivers, and swamps echoed to the calls of huge, long-dead animals. Indeed, the continents of North and South America were the last to be populated by humans and were the last to lose their diverse megafauna.

Further Reading: Harington, C.R. "Animal Life in the Ice Age." *Canadian Geographical Journal* 88 (1974): 38–43.

AMERICAN CHEETAH

Scientific name: *Miracinonyx trumani*
Scientific classification:
   Phylum: Chordata
   Class: Mammalia
   Order: Carnivora
   Family: Felidae
When did it become extinct? The American cheetah is thought to have become extinct around 10,000 years ago.
Where did it live? This cat was native to North America.

American Cheetah—Larger than the living cheetah, this North American predatory cat probably used speed to catch animals such as pronghorn. (Renata Cunha)

The pronghorn antelope (*Antilocapra americana*) of North America is one of the fastest land animals on the planet, able to reach speeds of 100 km per hour for short bursts and 40 to 50 km per hour over long distances. Why does it need such a turn of speed? There are no American predators that can sprint anywhere near fast enough to catch an adult pronghorn in a straight pursuit—well, there aren't any today. Some scientists believe that the pronghorns evolved to run so quickly as a way of evading an American cat that evolved along the same lines as the African cheetah (*Acinonyx jubatus*)—a slender feline built for speed. This was the American cheetah. The idea of a cheetahlike animal sprinting after pronghorns on the American Great Plains seems far-fetched, but prehistoric America was a very different place from the place we know today.

Remains of this sprinting cat are exceedingly rare, which is what you would expect for a light, slender-boned animal that was probably uncommon. With that said, the discoveries we have allow us to reconstruct what this animal may have looked like and how it may have lived. The bones of this animal were found in Natural Trap Cave, Wyoming—a big hole in the ground, where lots of prehistoric beasts met an unfortunate end—and Crypt Cave, Nevada. Up until the late 1970s, these bones were considered to be the remains of pumalike cats, but when some experts had a really close look at the bones, it was obvious that the cat in question was no puma. Like the modern-day cheetah, its skull had a very short muzzle, which gave it a rounded appearance, and the nasal cavities were very large. In the cheetah, these enlarged nasal cavities allow the animal to suck in big lungfuls of air during and after high-speed chases. The similarities also extend to the dentition as the modern cheetah has an interesting arrangement of cheek teeth, allowing the upper and lower sets to act like a single set of meat shears. As the skull of the extinct American cat had the same characteristics, we can assume that it had the same predatory lifestyle as the cheetah—a hunting strategy dependent on high-speed pursuit of fast-moving prey. Thanks to its big cheek shears, the African cheetah is one of the only cats that routinely eats bones, normally parts of ribs and vertebrae, and as the American cheetah's teeth are so similar, it may have done the same.

It is true to say that the skeletons of the long-dead American cheetah and the African cheetah are very similar, but there are some key differences, and one of the most obvious is size. On average, a fully grown African cheetah is around 67 kg. Using the skeleton of the American cheetah as a guide, this extinct animal may have been more like 80 kg. Also, the claws of the modern cheetah are completely nonretractable, a feature that gives the cat a good grip when it is pursuing prey (think of a human athlete wearing running spikes). The claws of the American cheetah could be fully retracted, which has led to the suggestion that this cat may not have been as specialized as the fast-running African feline we know today. The forelimbs of the American cheetah are also sturdier than today's cheetah, and they were sheathed in bigger muscles. Greater strength in the upper body, an interesting arrangement of the bones in the lower hind limbs, and retractile claws suggest that this animal may have been able to climb trees, something that today's cheetah definitely cannot do. However, these differences aside, so much of the American cheetah's skeleton is similar to the modern cheetah that it is very reasonable to assume these animals had very similar lifestyles.

How was the American cheetah related to the African cheetah? You would assume that being so similar, the American cheetah and the living African cheetah would be very closely related, and it has been argued that the American cheetah could have crossed the Bering land bridge into Asia, eventually arriving in Africa and spawning the cheetah we know today. However, nature is never that simple, and it is much more likely that these similarities arose due the process of convergent evolution—the phenomenon by which two unrelated species end up resembling one another because they adapt to similar circumstances.

Fortunately for the pronghorn antelope, the American cheetah died out around 10,000 years ago. Its extinction coincides with the disappearance of many North American mammals, but what factors ultimately led to the demise of this feline are more of a mystery. Climate change was obviously a factor, and the loss of some of its prey species may also have been important. It is possible that such a specialist cat really felt the squeeze of climate change and the effect it had on its environment. The puma, a generalist predator, is still with us today, but the American cheetah was more of a one-trick cat that survived by using speed to catch a small selection of prey animals. In today's big cats, we can see the price of extreme specialization, as the living cheetah is becoming increasingly endangered as its habitat is squeezed ever harder by human activities.

- The name "American cheetah" is often used to describe two extinct North American cats, the other being *M. inexpectatus*, which was a larger, and even more ancient species. In terms of appearance, this cat was halfway between the living cheetah and the living puma, and it may have been a more generalist predator than *M. trumani*.
- The cheetah and its prey (usually, gazelles, *Gazella* sp.) are often used to exemplify the concept of *evolutionary arms races*. In this case, the cheetah and the gazelle are locked in a struggle—if the cheetah evolves to run slightly faster, it will be able to catch more prey, weeding out the slower individuals from the population of gazelles; the surviving, faster gazelles pass on their fleet-footedness to their offspring, and eventually, these quicker individuals will predominate. So this process goes, with evolution continuously honing each species so that neither has the advantage for long.

+ Natural Trap Cave in Wyoming is a bell-shaped sinkhole at an altitude of around 1,500 m. Through a 4-m-wide hole at the surface, an unlucky animal would fall around 25 m to the cave floor. There is no route out of the cave once at the bottom, so if the unfortunate beast was not killed by the fall, it would have slowly starved. Over the millennia, lots of prehistoric and modern animals have stumbled into this hole, and it is now a site of extreme paleontological importance.

Further Reading: Adams, D. B. "The Cheetah: Native American." *Science* 205 (1979): 1155–58.

AMERICAN LION

American Lion—The American lion was substantially larger than the living lion. Bones from more than 100 individuals have been recovered from the Rancho La Brea asphalt deposits. (Renata Cunha)

Scientific name: *Panthera leo atrox*
Scientific classification:
    Phylum: Chordata
    Class: Mammalia
    Order: Carnivore
    Family: Felidae
When did it become extinct? The American lion became extinct around 10,000 years ago.
Where did it live? This cat was widespread in America, and its remains have been found from Alaska all the way down to Southern California. No remains have been found in the eastern United States or on the Florida peninsula.

The American lion is a very well known fossil animal. More than 100 specimens of this cat have been recovered from the asphalt deposits of Rancho La Brea alone, and disjointed bones and entire skeletons have been recovered from a host of other sites. All this material gives us a good idea of what this animal looked like as well as how it lived.

The bones of the American lion are very similar to the lion (*Panthera leo*) we know today, but scientists disagree on how these two animals are related. We do know that felines of lion proportions crossed into America via the Bering land bridge, and the American lion may simply be a subspecies of the living lion or possibly the same as the extinct European lion (*Panthera leo spelaea*), commonly known as the cave lion. Alternatively, the American lion may have been a distinct species and more similar, genetically, to the jaguar (*Panthera onca*). This extinct American cat was a big animal and one of the largest predators of the Americas, second only to the short-faced bears. It was around 25 percent larger than an average African lion, and it also had relatively longer legs.

We know this was a big, fearsome cat, but can ancient remains shed any light on how this feline lived? Is it possible to say whether the American lion was a social animal that lived and hunted in prides, as lions do today, or whether it was a solitary predator? Amazingly, there is some evidence to suggest that the American lion used teamwork to catch and subdue prey. This evidence is in the shape of a 36,000-year-old mummified bison that was found in Alaska by a gold prospector in 1979. Blue Babe, as this bison came to be known, has wounds that seem to be the work of two or three American lions. In the hide of this dead animal are the puncture wounds made by canine teeth and the characteristic slashes made by large feline claws. The only other animal capable of inflicting such wounds was the large scimitar cat, *Homotherium serum*, but a bite from this animal would have left a big tear in the skin, rather than puncture wounds. For some unknown reason, the lions that attacked this bison only ate part of the carcass before they were disturbed. We know the kill was made in winter as the bison had its winter coat and good stores of fat under its skin in preparation for the harsh conditions ahead. Perhaps some really bad weather closed in, forcing the lions to abandon their kill. Most tellingly of all, there was a large piece of American lion cheek tooth buried in the neck of the bison. Maybe the killers returned to the carcass after it had been frozen, and as they gnawed at the rigid flesh, one of them broke a tooth. The carcass was left for good and eventually covered by silt during the spring thaw, only to be unearthed by a high-pressure water hose 36,000 years later.

Finds like Blue Babe give us vivid glimpses of the how the American lion lived, and as with other extinct animals, the bones of the animal itself also tell many stories. Two specimens of the American lion from the Yukon show severe damage to the front of the lower jaw. The damage had healed, leaving large swellings on the mandible. We know that living lions are kicked in the face by struggling prey, and it seems that the American lion was also met with a hoof in the face when it was tackling the large herbivores of prehistoric North America. Not only did these cats get injured by their prey, but they also suffered from various diseases. One specimen from the Natural Trap Cave, Wyoming, has the telltale signs of osteoarthritis around the knee joint. This painful condition undoubtedly affected the ability of this individual to hunt effectively. Fast pursuits may have been impossible for it, so instead, it may have relied on scavenging, and perhaps it was the smell of decaying flesh that drew it to its death in the huge pitfall trap that is Natural Trap Cave.

Like all the other American megafauna, we will never know the exact cause of the demise of this cat. As a species, the American lion survived for many thousands of years, experiencing glaciations and warm interglacials, but like much of the American megafauna, it disappeared at the end of the last glaciation. Humans were spreading though North

America at this time, and as they hunted the prey of the American lion, this feline and humans were in direct competition. Various finds from around Europe show that prehistoric humans hunted lions, but it is doubtful whether direct human hunting could have led to the extinction of this cat. It is highly likely that this animal may have been better suited to the habitats and the colder conditions of the glaciations, rather than to the warm periods, and the pressure of climate change on its prey may have been amplified by human activity.

+ American lions were drawn to the Rancho La Brea because the sticky asphalt was a trap for all sorts of animals (see the "Extinction Insight" in chapter 4). The cats were attracted to the struggling animals, and they, too, became hopelessly stuck, eventually becoming entombed in the sticky tar. With this said, there are fewer American lions in the deposits than other predators such as saber tooth cats and dire wolves. Perhaps scavenging was only a last resort for the American lion, or maybe they were more wary of the potential dangers of tar pits.
+ You can see the mummified remains of Blue Babe in the University of Alaska Museum. It is known as Blue Babe because phosphorus in the bison's tissues reacted with iron in the soil to produce a white substance called vivianite. This mineral changes to a brilliant blue when it is exposed to the air.

Further Reading: Kurtén, B. "The Pleistocene Lion of Beringia." *Annales Zoologici Fennici* 22 (1985): 117–21.

WOOLLY RHINOCEROS

Woolly Rhinoceros—The woolly rhinoceros was widespread throughout northern Europe and Asia until the end of the last glaciation. (Phil Miller)

Scientific name: *Coelodonta antiquitatis*
Scientific classification:
   Phylum: Chordata
   Class: Mammalia

Order: Perissodactyla

Family: Rhinocerotidae

When did it become extinct? This rhinoceros is thought to have survived until around 10,000 years ago.

Where did it live? This was probably the most widespread rhinoceros of all time as its remains have been found all the way from Scotland to South Korea, and to Spain in the south of Europe.

The world is still in the grip of a cold period, and it has been for the last 40 million years or so. Around 3 million years ago, this cold period intensified, and huge ice sheets spread across much of the Northern Hemisphere. From then until now, the ice sheets have waxed and waned in fairly regular cycles played out over 40,000- to 100,000-year cycles. During this intensification, life had to adapt, move, or go extinct. The rhinoceri, with the thermal inertia afforded by their big, heavy bodies, were well placed to take advantage of these icy conditions, and the fossil record shows that they positively embraced the ice age and expanded their range to cover much of the Northern Hemisphere. This success was mainly due to one species: the woolly rhinoceros.

The woolly rhinoceros was about the same size as the biggest living rhinoceri, the white and Indian rhino, but thanks to its shaggy coat, it probably looked a lot more imposing. This ice age brute was around 1.8 m tall and 3.5 m long, and it probably weighed in the region of 3 tonnes. On its head were two horns, the longest of which was around 1 m. As its name suggests, the woolly rhinoceros was completely clothed in thick fur, and this pelage, together with a thick layer of fat beneath its skin, helped to insulate its body from the cold. Our ancestors were obviously well aware of the woolly rhinoceros as it has been depicted numerous times in European cave paintings. Some of these cave paintings appear to show the woolly rhinoceros with a dark-colored band of fur around its middle. Whether this was artistic license on behalf of the prehistoric painter or the genuine appearance of the animal is impossible to know, but these images do give us a tantalizing glimpse of the world through the eyes of our ancient ancestors.

The realm of this great, furry beast was the tundra and steppe that extended out in front of the immense ice sheets that capped the Northern Hemisphere. This was a harsh environment, but lots of animals appeared to have thrived in these cold conditions. Like the rest of its kind, the woolly rhinoceros was a herbivore, but was it a browser or grazer? This question has divided paleontologists for years, but it is very likely that this giant was a grazer. The woolly rhinoceros's neck muscles were very powerful, which is just what you would expect for an animal that had to tear mouthfuls of grass from the ground. Also, ancient, buried pollen from long-dead flowering plants can tell us a lot about the ancient earth, and in the places where the woolly rhinoceros was found, the most common plants were grasses and sedges. On the icy steppe and tundra, grass was covered for some of the year by snow, and it seems these big plant eaters got at their food by using their long horn to brush away the snow. The woolly rhinoceros horns that have been unearthed show abrasive wear on their outside edges, indicating that they were probably swung to and fro along the ground to sweep the snow from the grass. Grasses and sedges may have been abundant on the ice age steppe and tundra, but these plants are far from easy to digest. Every cell in a blade of grass is packed with proteins, fats, and carbohydrates, but these nutrients are difficult to get at because the cell is encased in a wall of cellulose. Mammals like the woolly rhinoceros can only digest grass with the help of symbiotic bacteria. In rhinoceri and other herbivores, such as horses and rabbits, these bacteria are found in the back end of the animal's gut, and here they digest the tough cellulose cell walls of the plants to release the contents. Some of these nutrients are used by the bacteria, and some are absorbed by the herbivore. This is quite an inefficient process, so the woolly rhinoceros spent a lot of time each day eating to supply its considerable bulk with sufficient energy.

The oldest remains of the woolly rhinoceros are around 350,000 years old and it is possible that even older bones may be lying in the ground awaiting discovery. During its time on earth, the woolly rhinoceros experienced a number of global cooling and warming events, and its populations probably expanded and contracted, reflecting the movement of the great ice sheets. With this in mind, could climate change alone have led to the extinction of this animal? No is the likely answer, and it was probably a combination of factors that led to the extinction of this megaherbivore. What about hunting? We know that Neanderthals and our ancestors hunted this animal. To them, a fully grown woolly rhinoceros was a massive source of meat, fat, bone, fur, and leather, and the killing of such a large, dangerous animal was probably seen as a very risky undertaking—an act that the warriors within the tribe may have used to demonstrate their bravery. Climate change definitely squeezed the populations of this animal, especially the warmer cycles, and with their populations under pressure, human hunting may have been sufficient to kill them off completely or reduce their numbers to a point from which recovery was impossible.

- In 1929, in Staruni, Ukraine, an amazingly well preserved female woolly rhinoceros was discovered. Apart from the fur and hooves, the carcass was complete. It had come to rest in oil- and salt-rich mud, and these conditions had prevented bacterial decay. It is currently on display in the Krakow Museum of Zoology, Poland.
- It has been suggested that as the woolly rhinoceros was such a large, dangerous animal, prehistoric humans may have hunted it using traps, instead of facing it directly and risking a goring on the end of its impressive horn. Perhaps ancient hunters drew them to excavated pits or deep mud. Once trapped, the struggling rhinoceri could have been safely killed with spears.
- The closet living relative of the woolly rhinoceros is thought to be the rare Sumatran rhinoceros (*Dicerorhinus sumatrensis*). This shy, rarely seen animal is the smallest of the living rhinoceri, and bizarrely, its whole body is thinly covered in reddish fur.

Further Reading: Orlando, L., J.A. Leonard, V. Laudet, C. Guerin, C. and Hänni. "Ancient DNA Analysis Reveals Woolly Rhino Evolutionary Relationships." *Molecular Phylogenetics and Evolution* 28 (2003): 76–90.

LITOPTERN

Scientific name: *Macrauchenia patachonia*
Scientific classification:
  Phylum: Chordata
  Class: Mammalia

Litoptern—The litopterns were an unusual group of herbivorous mammals found throughout South America. The species depicted here was the last of their kind. (Renata Cunha)

Order: Litopterna
Family: Macraucheniidae
When did it become extinct? This species of litoptern became extinct around 10,000 years ago.
Where did it live? The litopterns were found only in South America.

In 1834, the young Charles Darwin discovered the foot bones of an extinct herbivorous mammal in Patagonia. Initially, these bones were thought to have once taken the weight of a giant llamalike animal, but it was later realized that they belonged to a very different creature. Most of the large plant-eating mammals that have wandered the earth for the last few thousand years can be divided into two major groups: the odd-toed ungulates (perrisodactyls)—animals like horses and rhinoceri—and the even-toed ungulates (artiodactyls), a group that includes deer, cattle, and so on. In the years following Darwin's discovery, more finds came to light, and it slowly became clear that up until about 10,000 years ago, South America had its own large plant-eating mammals, and they were unique—quite different from the odd-toed and even-toed ungulates.

One group of these unique, South American herbivores was the litopterns. The last of the litopterns to become extinct looked like a stocky, humpless camel with thick legs. With a shoulder height of 1.5 m and a body length of 3 m, this litoptern was one of the larger South American mammals. The nasal openings on the skull of this animal are very near the top of the head, which has led paleontologists to believe that they probably had a short trunk. We can't be sure what a stubby trunk was for, but it may have been used to grasp low branches and to pull them within reach of its mouth, in the same way that a giraffe uses its long prehensile tongue to grab vegetation on high branches. The largest of the litopterns had a three-toed, flat-footed stance, but some of the more lightly built species had slim legs that ended in a single toe. At the center of these slim legs were strong bones and flexible joints—the hallmarks of fleet-footed animals that evade their enemies with speed and maneuverability.

These odd plant eaters needed some way of evading danger as prehistoric South America was home to lots of fearsome meat eaters. There were fearsome felines, killer marsupials, and terrifying birds, so the litopterns must have always been on the lookout for danger as any one of these predators was quite an adversary. Even an adult of the biggest of these bizarre herbivores was no match for the most powerful saber tooth cat that has ever lived (*Smilodon populator*—see the entry "Saber Tooth Cat" earlier in this chapter). The soft underside of the litoptern's long neck was probably a very attractive target for the saber tooth cat, and it is likely that they were commonly killed and eaten by these formidable felines. The larger terror birds must have been quite capable of killing the smaller litopterns as well as juveniles of the larger species.

Speed gave the smaller litopterns some protection from predators, but their most effective defense was probably strength in numbers, and it is very likely that these extinct animals lived in herds in the same way as living herbivorous mammals. It makes sense for any large animal with lots of enemies to live in herds as there will always be several pairs of eyes on the lookout for danger. Almost all predators rely heavily on the element of surprise, and without this, they stand little chance of making a successful kill. The litopterns used their keen senses of sight, smell, and hearing to alert the herd to danger. When a predator did launch an attack against a herd of these animals, it is likely that they singled out the young, old, and sick animals as a healthy adult litoptern must have been very difficult to catch.

With numerous predators all wanting to get their teeth, claws, and beaks into the succulent flesh of litopterns, life on the grasslands of South America must have been very difficult for these herbivores, and about 2.5 million years ago, something happened that made things even more difficult. The Great American Interchange saw all sorts of wildlife stream into South America from the north. Some of these creatures competed with litopterns for food, and others hunted them. All these new challenges were played out within the backdrop of a changing world. Sea levels were falling and the global climate was becoming drier and cooler—bad news for trees, the likely favored food of the litopterns. The times following the Great American Interchange must have been very tough for these odd ungulates. Their habitat was disappearing, strange animals from the north competed with them for the food that was left, and a number of cat species, also immigrants from the north, and deadly in tooth and claw, were well equipped to hunt the remaining litopterns. Long after the Great American Interchange reached its peak, humans spread throughout South America, and they, too, must have hunted the remaining populations of litopterns, which, by that point, must have been reduced to a shadow of their former strength. Squeezed from all sides, these unique, plant-eating mammals eventually became extinct around 10,000 years ago.

+ Darwin's discovery of the first litoptern fossils was made during his voyage on HMS *Beagle*. The ship stopped for some time in Patagonia, allowing Darwin to explore these lands, and it was then that he discovered and excavated the bones of several extinct South American mammals.

+ The name "litoptern" means "simple ankle" and refers to the bone structure of the ankle of these animals as the articulation of the bones is not as complex as in other large herbivorous mammals.
+ Fossils of *Macrauchenia patachonia* have been found in Patagonia all the way up to Bolivia, so this was a very widespread species. Many litoptern fossils have been discovered in the Lujan formation, near Buenos Aires in Argentina.
+ The placental mammalian predators that moved into South America during the Great American Interchange included familiar animals like the jaguar and puma.

Further Reading: MacFadden, B.J. "Extinct Mammalian Biodiversity of the Ancient New World Tropics." *Trends in Ecology & Evolution* 21 (2006): 157–65; Reguero, M.A., S.A. Marenssi, and S.N. Santillana. "Antarctic Peninsula and South America (Patagonia) Paleogene Terrestrial Faunas and Environments: Biogeographic Relationships." *Palaeogeography, Palaeoclimatology, Palaeoecology* 179 (2002): 189–210; Tonni, E.P., A.L. Cione, and A.J. Figini. "Predominance of Arid Climates Indicated by Mammals in the Pampas of Argentina during the Late Pleistocene and Holocene." *Palaeogeography, Palaeoclimatology, Palaeoecology* 147 (1999): 257–81.

DIRE WOLF

Scientific name: *Canis dirus*
Scientific classification:
    Phylum: Chordata
    Class: Mammalia
    Order: Carnivora
    Family: Canidae
When did it become extinct? It is thought that the dire wolf became extinct sometime between 8,000 and 16,000 years ago.
Where did it live? Dire wolf fossils have been found in North, Central, and South America.

If you are alone in the woods, the howl of the wolf must be one of the most unnerving sounds you can hear. Our fear of this sound is an old one that extends all the way back to prehistory, when the human race was more closely tied to the natural world. Thousands of years ago, the woods and wide open spaces of North America echoed to not one species of wolf, but two. One of these, the gray wolf (*Canis lupus*), is still with us, although its range is only a fraction of what it used to be. The second species, the dire wolf, became extinct thousands of years ago. The dire wolf is the largest, non-domesticated canine that has ever lived, and thanks to the huge numbers of fossils that have been found, one of the best known of all recently extinct carnivores.

On average, the dire wolf was only slightly larger than the biggest gray wolf, at around 65 kg in weight and 1.5 m in length (Great Danes are around the same weight, but they are taller and more slender). It had a slightly heavier build than the gray wolf and a relatively larger head. Interestingly, its legs were relatively shorter than those of the gray wolf. If you compare the skeleton of a dire wolf to that of a gray wolf, the biggest differences are the skull and teeth. The skull of the dire wolf not only looks a lot heavier than the gray wolf's skull, it also contains more impressive teeth.

Dire Wolf—The dire wolf was more heavily built than the living wolf. The remains of more than 1,600 dire wolves have been unearthed from the Rancho La Brea asphalt deposits. (Renata Cunha)

The differences in the skeleton of the dire wolf compared with the living wolf give us some clues to how this extinct dog may have lived. The gray wolf is built for stamina and long-distance pursuit. It has long legs, a narrow chest, and a long, flexible back, enabling it to cover long distances in bounding strides. The dire wolf, on the other hand, had relatively shorter legs, and this has led some scientists to suggest that it was not much of a long-distance runner, although it could have undoubtedly burst into a sprint when the need arose.

The teeth of the dire wolf are more robust than those of the living wolf, and the well-developed cheek teeth were probably used to crack the bones of carcasses. Unfortunately, we can never know for sure how the dire wolf lived, but like the living gray wolf, it was very probably an opportunist, switching between active predation and scavenging, depending on the situation. Perhaps the dire wolf was a capable predator like its living relative, but with more of a preference for large, slow-moving prey. To supplement hunting, the dire wolf probably scavenged whenever possible. Thousands of years ago, North America was dominated by megafauna—large mammals, including the mammoths, mastodons, giant deer, and many other species—almost all of which are now extinct. Instead of running to escape their enemies, many of these animals depended on their size for protection, and therefore speed and stamina may have given the dire wolf little or no advantage in hunting the herbivores of the megafauna. However, a powerful bite and more robust body made it easier for these wolves to hang on to, and eventually subdue, large prey animals.

The dire wolf may have been more heavily built than the gray wolf, but its brain was actually smaller in absolute terms. We know the gray wolf is a very sociable animal, living in tightly knit packs. Hunting as a team allows the gray wolf to catch and kill prey that would be far too big for a single wolf to bring down. As a rule of thumb, animals with an elaborate social behavior have a relatively bigger brain than solitary animals. So does the smaller brain of the dire wolf mean it was less capable, socially, than its extant relative? We don't know for sure, but the discovery of huge numbers of dire wolf skeletons in the Rancho La Brea asphalt deposits in Los Angeles, together with the fact that all larger surviving canids are social, suggests that the dire wolf lived in packs.

How come the Americas no longer echo to the howls of two wolf species? What happened to the dire wolf? It seems this canine was another casualty of the megafauna collapse that swept through the Americas between 16,000 and 8,000 years ago. The earth's climate was going through some huge changes as the last ice age was coming to an end, and humans were spreading into the New World along dispersal routes that took them across the Bering land bridge and farther south and east by land and sea. The herbivorous mammal megafauna of these continents appears to have dwindled and vanished in the face of these changes, but we shall never know the degree to which human hunting caused these declines. Changing habitats and disappearing prey, especially loss of large herbivores due to ecosystem change as well as hunting by humans, eventually impacted the number of predators. The heavily built dire wolf, with its probable preference for large prey, felt the changes more than its relative, the gray wolf, and eventually died out.

+ The Rancho La Brea asphalt deposits of Los Angeles, California, have yielded the remains of more than 1,600 dire wolves—one of the most common predators at the site (see the "Extinction Insight" in chapter 4). In the site museum, there is an entire display wall made up of 450 dire wolf skulls.
+ The wolves in the asphalt deposits were trapped over a period of thousands of years and were attracted to prey animals that had also gotten themselves trapped in the sticky goo. The wolves probably pounced on the unfortunate prey, and they, too, found themselves stuck, with nothing but a slow, miserable death ahead of them.
+ How come so many dire wolves met a sticky end in the asphalt deposits of Rancho La Brea? They must have been very numerous animals, very stupid, or overly aggressive. It seems that some predators were aware of the dangers of the pits, or at least were repelled by the odor of the tar. Whatever the reason, the treasure trove of dire wolf remains from Rancho La Brea gives us an unsurpassed record of the appearance and life of this animal.

Further Reading: Leonard, J.A., C. Vila, K. Fox-Dobbs, P.L. Koch, R.K. Wayne, and B. Van Valkenburgh. "Megafaunal Extinctions and the Disappearance of a Specialized Wolf Ecomorph." *Current Biology* 17 (2007): 1146–50.

CAVE BEAR

Scientific name: *Ursus spelaeus*
Scientific classification:
  Phylum: Chordata

Cave Bear—The cave bear's steep forehead is clearly visible in this picture. Much of the space inside the forehead is taken up by the structures that supported the nasal tissue. This probably gave the bear an excellent sense of smell. (Ross Piper)

Cave Bear—Our ancestors would have frequently encountered the cave bear as they sought refuge in caves. (Phil Miller)

Class: Mammalia
Order: Carnivora
Family: Ursidae

When did it become extinct? Although they disappeared from many areas of Eurasia as early as 20,000–30,000 years ago, the cave bear finally became extinct around 10,000 years ago.

Where did it live? The remains of the cave bear have been found throughout Europe, to Russia in the east, Spain in the south, and France in the northwest. They may even have reached Britain.

Of all the animals that have become extinct during the last 10,000 years or so, the remains of the cave bear are among the most numerous. In Dragon's Cave, near Mixnitz in Austria, the remains of around 50,000 cave bears have been found. Indeed, almost every cave in central Europe with an entrance big enough to permit the entry of a large animal will have, at some point, played host to this extinct bear. Suitable caves were used by generation upon generation of these bears over hundreds of thousands of years, and where the passages are quite narrow, the walls have been polished by the comings and goings of countless furry bodies over the ages.

Superficially, the cave bear was very similar to the living brown bear, but it was undoubtedly a distinct species. It was much bigger than a European brown bear (*Ursus arctos*), and some adult males, fat from feeding in preparation for the winter hibernation, probably weighed in the region of 400 kg. The cave bear also had a relatively large head and short, powerful limbs, while the brown bear has a leggier appearance. The most notable feature of the cave bear's skeleton is its large, domed skull, which has a characteristically steep forehead. Unfortunately for this extinct bear, this vaulted cranium did not house an enlarged brain. Cutting one of these ancient skulls in half shows that the braincase of the cave bear was no bigger than that of the brown bear, and much of the extra space is actually taken up by air spaces and all the elaborate structures that gave the cave bear a very acute sense of smell. The living bears are renowned for their keen sense of smell, but it seems the cave bear could probably outperform any bear, living or dead, in tracking scents.

As with the bones of the giant short-faced bear (see the entry on this animal in chapter 6), chemical analyses of the remains of the cave bear have revealed some interesting things about the life of this animal. Most living bears are opportunistic omnivores that eat whatever, whenever they can find it. Apparently, the cave bear was predominantly a herbivore that probably fueled its bulk with all manner of succulent leaves, bark, roots, tubers, fruits, nuts, and seeds. Although the teeth of the cave bear are undoubtedly those of a generalist, they have lost some of the carnivorous edge seen in the dentition of the living brown bear. Its molars were absolutely massive, perfect for crushing and pulverizing tough plant foods. Although the cave bear may have been very keen on plant food, it is very unlikely it turned its sensitive nose up at the chance of consuming meat when it was easily available such as from an abandoned carcass. There are even bones from some localities that suggest that, at least in some places, the bears were mostly feeding on meat.

Because so many remains of the cave bear have been found over the years, we have a very good picture of what ailments these animals suffered from. The skeletons of many cave bears show signs of osteoarthritis, and in severe cases, the vertebrae of the spine have fused together or elaborate outgrowths of bone have sprouted from the limbs. Severe cases of this disease made the suffering animal lame. In many skeletons, there are the telltale signs of severe dental wear and disease, and in the sinuses of the heads of a few individuals, there are the large pockmarks of bone-eating bacterial infections. Although cave bears were robust animals, they broke their bones in falls and fights, and in many cases, the bones were misaligned when they knitted back together, leaving the poor animal crippled, but apparently still able to survive. Some individuals also suffered from rickets, a disease caused by a lack of vitamin D and which results in bone deformities such as bowed limbs. Bears, like humans, synthesize vitamin D in their skin in the presence of sunlight, and as cave bears were forced to see out the harsh winter by hibernating in rocky refuges, they often didn't produce enough vitamin D. Rickets was particularly common in bears living at high altitudes due to the short ice age summer season in the high mountains. These high mountain bears were forced to spend more time in their caves than bears living at lower elevations. Perhaps the most bizarre affliction is the injury sustained by some unfortunate male bears. All male bears, and many other male mammals, for that matter, have a bone in their penis called the baculum. In some male cave bears, this bone was broken, but exactly how it was fractured is a mystery. Was it broken when a mating male bear was fending off other potential suitors, or was it stepped on in a dark cave long after the bear died?

This extinct bear's predilection for caves must have brought it into direct competition with prehistoric humans, who prized these places as refuges from the elements and predators. Our Pleistocene ancestors definitely knew of the cave bear and even depicted it in various paintings, which can be seen in a number of caves throughout Europe. One very interesting painting shows a cave bear that seems to be bristling with spears, blood gushing from its mouth. Early cave bear finds suggested to some experts that our ancestors may have revered these animals. High up in the Swiss Alps, Drachenloch Cave was reported to contain what appeared to be the oldest stone structure of religious significance anywhere in the world. Attributed to Neanderthals who lived 70,000 years ago, this case was the basis for the so-called Cave Bear Cult. One cave bear skull was even found with a cave bear femur twisted behind the cheekbone and was considered to be the work of human hands. The famous paleontologist Björn Kurtén challenged this view and suggested that the peculiar arrangements of bones could have been produced by chance as other bears shoved old bones around the cave floor, preparing their winter retreat.

These and other finds give us a fascinating glimpse of how early humans and long-dead animals interacted in a very different world. Some cave bear bones have been found bearing the scorches from fire and the cut marks from stone tools. These show that prehistoric humans hunted the cave bear. But did they drive it into extinction? The cave bear survived hundreds of thousands of years of oscillating climatic conditions and changing habitat, and in the very harsh glacial periods, the species may have been reduced to small populations that managed to cling to survival in sheltered valleys. Perhaps it was human hunting in combination with the pressure of a changing climate that led to the demise of these interesting mammals.

+ During the age of discovery, when gentlemen scholars started to probe the prehistory of the earth, it was thought that the cave bear fell into two distinct groups: dwarves and giants. In actual fact, the difference in size of the cave bear skeletons was due to sexual dimorphism: adult male cave bears could weigh twice as much as females.
+ In some caves, there are deep scratches in the walls, which were almost certainly left by the claws of a cave bear. Were they marking their bedding areas, or were they trapped in a rock fall? It is thought the former is more likely.
+ A complete flint weapon tip was found in a cave bear skull discovered near Brno in the Czech Republic, indicating that a human hunter was trying to kill the animal at very close quarters. Hunting a cave bear must have been a very dangerous business. They may have been herbivorous, but they were immensely strong and probably very easily angered. A large, rearing cave bear was at least 3 m tall, and a swipe from one of its massive paws would have easily snapped the neck of a human assailant.

Further Reading: Kurten, B. *The Cave Bear Story*. New York: Columbia University Press, 1976.

SICILIAN DWARF ELEPHANT

Scientific name: *Elephas falconeri*
Scientific classification:
    Phylum: Chordata
    Class: Mammalia
    Order: Proboscidea
    Family: Elephantidae
When did it become extinct? This elephant became extinct around 10,000 years ago.
Where did it live? This animal was endemic to some of the Mediterranean islands and many remains have been found in Sicily.

Today, the multitude of islands that dot the Mediterranean are where many Europeans choose to spend their summer vacations. Long before these islands became destinations for vacationers, they played host to various key events in human history. Some of the earliest civilizations had their beginnings on these islands, but if we travel even further back, to a time before modern humans started to leave Africa, these islands supported their own endemic animals, almost all of which are sadly extinct.

Sicilian Dwarf Elephant—The Sicilian dwarf elephant, only about 1 m at the shoulder when fully grown, once roamed around the island of Sicily. (Phil Miller)

Sicily is one of the more well known Mediterranean islands, important throughout antiquity because of its strategic location. Up until around 11,000 years ago, Sicily was free of humans, and a number of mammals had taken up residence on the island and evolved into distinct species. One of the most bizarre inhabitants of the prehistoric Sicily was the dwarf elephant. As with all terrestrial island mammals, we can never be sure how the ancestors of the dwarf elephant reached Sicily, but they could have swam or crossed via a temporary land bridge that was revealed when sea levels were much lower. Elephants will take to the water without hesitation, and they can even use their long trunk as a snorkel (there are many reliable reports of Indian elephants, *Elephas maximus*, being sighted several kilometers out at sea). It is thought that the dwarf elephant evolved from the straight-tusked elephant (*Elephas antiquus*), an inhabitant of Europe up until around 11,500 years ago. Searching for new areas of habitat, the elephants took to the water or crossed a land bridge, eventually reaching Sicily.

Even though Sicily is one of the largest islands in the Mediterranean, it is a small landmass, and a straight-tusked elephant, at around 10 tonnes, is a huge animal with a big appetite. Living, fully grown elephants require about 200 kg of food every day to survive. This can be sustained on the mainland, where the animals can move to new areas of habitat, but the vegetation on an isolated landmass would quickly be exhausted by the immense appetites of these creatures. To adapt to life on their new island, something strange happened to the straight-tusked elephants: they began to shrink. Generation after generation, the elephants diminished in size to adapt to the limited food resources on Sicily. This phenomenon is known as the island rule, and it can be seen all over the world, wherever animals take up residence on isolated islands. The shrinkage of the Sicilian elephants is identical to what happened to the hominids who made it to Flores in Indonesia (see the entry "Flores Human" in chapter 6).

It is impossible to say how long the dwarfing process took, but it was probably very quick in evolutionary terms as failure to adapt to new surroundings swiftly leads to extinction. After thousands of years of gradual shrinkage, the Sicilian elephant was a fraction of the size of its ancestors. In life, it probably weighed around 100 kg, about 1 percent the size of a large straight-tusked elephant. Although food is a limiting factor on small islands that can lead to dwarfing, the lack of predators and competition are also important. Large size is an excellent defense against predators, but on Sicily, where predators were notable by their absence, there was no advantage in being big. Large size can evolve in a species due to competition because in some ways, a larger body is more efficient than a small one. In the absence of this pressure, the species may shrink as a lot of resources and time are needed to grow to a large size.

Exactly how these tiny, extinct elephants lived will never be known, but an animal only slightly heavier than a pig had quite a different life than its enormous ancestors. Straight-tusked elephants on the mainland were able to feed on tree leaves and other lofty plant matter, even uprooting whole trees they liked the look of, but the Sicilian dwarf may have fed on low-growing vegetation, perhaps using its trunk to bring low-growing vegetation within reach of its mouth. Bushes and low-growing plants, such as grasses, probably featured prominently in the diet of this Mediterranean dwarf. Elephant digestion is very inefficient, and around 60 percent of all food leaves the gut of these animals undigested. Even though the dwarf elephant was only a fraction of the size of its ancestor, it may still have needed several kilograms of food every day. If this was the case, such feeding demands on an island that had not previously known any large herbivores must have had a huge effect. The feeding activities of the elephants and the damage they caused as they were trudging around the island may have reshaped the whole ecosystem of Sicily. It is possible that the number and diversity of plants on the island probably underwent dramatic changes as the elephant population grew to its maximum. Some plants may have suffered due to disturbance and elephant feeding, whereas others may have benefited from an increase in glades and other open areas and the valuable influx of nutrients that large herbivore dung provides.

The dwarf elephants survived on Sicily for hundreds of thousands of years, but like the straight-tusked elephants before them, humans on the mainland were searching for new places to live. They, too, set off across the Mediterranean, in boats and traversing land bridges, hoping to find new lands. They found Sicily and its dwarf elephants around 11,000 years ago. Because the dwarf elephants had been isolated for so long, they lacked the innate fear of humans possessed by most mammals. Elephants are curious, intelligent creatures, and they probably investigated the first humans they saw. A 100-kg animal could feed a tribe of hungry humans for many, many days, and the dwarf elephant's lack of fear made it very easy to hunt. Sicily could have supported no more than a few hundred dwarf elephants, and this small population was probably wiped out in a few decades.

- ◆ The islands of the Mediterranean are part of the continental crust, rather than being created relatively recently by volcanic activity. Therefore they have been around for a long time—isolated outposts in the azure waters of the Mediterranean.

♦ The migration of straight-tusked elephants to Sicily was not an isolated event. Many islands in the Mediterranean had their own species of dwarf elephant, which descended from the big, straight-tusked elephant that swam across from the mainland or traversed a temporary land bridge or series of small islands. The Cypriot dwarf elephant (*Elephas cypriotesi*) was about twice the size of the Sicilian species, but it was still very much smaller than its ancestors.

Further Reading: Masseti, M. "Did Endemic Dwarf Elephants Survive on Mediterranean Islands up to Protohistorical Times?" In *The World of Elephants—International Congress*, 402–6. Rome, 2001; Palombo, M. "Endemic Elephants of the Mediterranean Islands: Knowledge, Problems and Perspectives." In *The World of Elephants—International Congress*, 486–91. Rome, 2001; Masseti, M., and M.R. Palombo. "How Can Endemic Proboscideans Help Us Understand the 'Island Rule'? A Case Study of Mediterranean Islands." *Quaternary International* 169/170 (2007): 105–24.

MERRIAM'S TERATORN

Merriam's Teratorn—Merriam's teratorn was larger than the living condors, and its remains have been found throughout much of the United States. (Renata Cunha)

Scientific name: *Teratornis merriami*
Scientific classification:
    Phylum: Chordata
    Class: Aves
    Order: Ciconiiformes
    Family: Teratornithidae
When did it become extinct? This bird died out around 10,000 years ago.
Where did it live? The remains of this bird have been found in various locations in North America, including California, southern Nevada, Arizona, and Florida.

The Rancho La Brea asphalt deposits in California have yielded a huge number of fossils, the remains of animals that became entombed in sticky tar between 8,000 and 38,000 years ago (see the "Extinction Insight" in chapter 4). Bird fossils, rare elsewhere because they are so fragile, have been found in abundance at Rancho La Brea. This one rich deposit of fossils gives us an excellent idea of what birds lived in that corner of California all those millennia ago. Some of these birds are still with us today, while others are only known from their bones. One of most remarkable extinct birds from the deposits is Merriam's teratorn, a relative of the colossal magnificent teratorn of South America (see the entry "Magnificent Teratorn" in chapter 6) and the living condor species. It is the most well known of all the teratorn species as the bones of more than 100 individuals have been recovered from Rancho La Brea.

Merriam's teratorn was diminutive compared to the magnificent teratorn, but by today's standards, it was a giant. With a wingspan of around 3.8 m and weighing in at about 15 kg, the closest comparable living bird is the Andean condor, one of the Merriam's teratorn's closest living relatives. The Andean condor is a scavenging bird of prey that uses its immense wingspan to soar effortlessly on the thermals that rise into the air around the flanks of mountains as the sun warms the ground. High in the air, the condor can scan the ground below for its favorite food: carrion. When Merriam's teratorn was initially described in 1909, it was assumed that the living bird was primarily a scavenger due to its similarities with the living condor. Many decades later, paleontologists closely studied the skulls of this extinct bird and came to the conclusion that in life, Merriam's teratorn was an active predator that spent a good deal of time on the ground, prowling areas of short vegetation for small mammals and other delicious morsels. Other experts have compared the skull of Merriam's teratorn to skulls from a number of other predatory birds, including living and extinct species. These comparisons indicate that the extinct giant may have been a specialist fish predator. If this was the case, it may have had trouble plucking prey from the surface of the water with its feet as they don't seem to be up to the job of grasping a slippery fish. Some birds (e.g., frigate birds, *Fregata* sp., and some terns) pluck fish from the water with their beaks, and perhaps this is what Merriam's teratorn did as it was gliding just above the surface of the water. If Merriam's teratorn specialized in taking fish on the wing from inshore waters, then its abilities in the air must have been staggering. With a wingspan approaching 4 m, any wrong move just above the water's surface must have ended in a very wet teratorn, and one that probably could not take off again.

If this giant bird was able to pluck fish from the surface of calm, inshore waters, why have so many specimens been found in the asphalt deposits of Rancho La Brea? Birds of prey were drawn to Rancho La Brea for one thing: carrion. Animals of every description met a slow and grisly end in these sticky tar pits, and the larger ones, in their struggles to free themselves, must have attracted predators from far and wide. Saber tooth cats came to try their luck, as did dire wolves and a range of other large predators. Many of these also became trapped until the sticky goo was a banquet of dead and dying animals, just the sort of thing to appeal to scavengers. Merriam's teratorn and a host of other scavenging birds, including condors, eagles, and ravens, probably perched in trees near the edge of the tar pits waiting for the final, futile struggles of a large mammal. With the poor animal still alive, the scavengers descended and perched on the hulking brute, tearing at the tar-matted hide with their sharp beaks. Just like today, squabbles among scavenging animals were commonplace, and the teratorns probably jostled for space on the carcass until one of them lost its footing and ended up in the tar itself.

In terms of behavior, the closest comparable living bird to Merriam's teratorn may be the bald eagle (*Haliaeetus leucocephalus*), a fish specialist with no qualms about eating carrion from a large carcass. Perhaps Merriam's teratorn, like the bald eagle, was primarily a fish eater but was occasionally drawn to the bounteous supply of carrion at Rancho La Brea, which, back then, was very close to the coastline.

For hundreds of thousands of years, this immense bird graced the skies of North America, and it was undoubtedly known to Amerindians, who apparently hunted it. Although Merriam's teratorn may have had a similar lifestyle to the living bald eagle, it must have been heavily dependent on carrion, particularly the carcasses of large mammals, as its demise coincides with the disappearance of the large North American animals. With suitable carcasses becoming scarcer and scarcer and humans hunting them for food, the long-lived but slow-breeding Merriam's teratorn was doomed, and sadly, it died out.

- Merriam's teratorn was the largest flying bird ever seen alive by humans, and it is very possible that this extinct giant could be the inspiration for the mythological thunderbird. In Amerindian stories, this enormous bird is said to cause thunder by flapping its wings, and its likeness is often seen surmounting totem poles. Perhaps the teratorn obtained immortality in the oral traditions and stories of the Amerindians as a folk memory.

- A large, dead mammal can supply a large variety of scavenging animals with a huge amount of food, but there is always a pecking order at a carcass. Like the living lappet-faced vulture (*Torgos tracheliotus*), Merriam's teratorn may have been the only scavenging bird capable of tearing through the thick skin and tough muscle of a large, fresh carcass. Just like today, the lesser scavengers may have had to wait until the teratorn ate its fill.

Further Reading: Campbell, K. E., Jr., and E. P. Tonni. "Size and Locomotion in Teratorns." *The Auk* 100 (1983): 390–403; Hertel, F. "Ecomorphological Indicators of Feeding Behavior in Recent and Fossil Raptors." *The Auk* 112 (1995): 890–903; Harris, J. M., and G. T. Jefferson, eds. "Rancho La Brea: Treasures of the Tar Pits." *Los Angeles City Museum Science Series* 36 (1985).

Extinction Insight: Ice Ages

It took a very long time indeed for humans to determine that the earth is far from the stable home we think it is. Changes in the geometry of the oval orbit along which the earth circles the sun, changes in the earth's tilt over time, changes in the way in which the earth progresses along this orbit, and other, as yet unknown factors all contribute to what can only be described as a very variable climate, both in the short and long term.

For immense stretches of time, the earth has been a hothouse with no trace of ice anywhere, a colossal ball of ice, and every variation in between. Currently we consider ourselves to be in the middle of a pleasant, balmy period, but in actual fact, the earth is locked in the grip of an ice age, and it has been for the last 2 million years. Phrases like "since the last ice age ended" are a dime a dozen, but the truth of the matter is that we are merely in what is known as an interglacial, a warm period sandwiched between much colder, glacial periods. The earth actually first entered this cold phase about 40 million years ago, when the Antarctic ice sheet began to form, but it is only in the last 1.6 million years that the earth's climate has oscillated between long, cold periods (glacials) and short, warmer periods (interglacials).

Scientists have worked out that over the last 1.6 million years, there have been at least seven of these glacial-interglacial cycles, and possibly many more. How can scientists know what the climate was like hundreds of thousands of years ago, when it is still impossible to forecast, with 100 percent accuracy, the weather tomorrow? Every single day, an enduring record of the earth's climate is stored away on the seafloor or in ice sheets. The record deposited on the seafloor is not in words or numbers, but is codified in the remains of microscopic, planktonic organisms

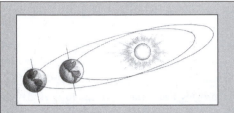

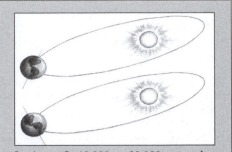

Ice Ages—In 100,000-year cycles, the eccentricity of earth's orbit changes. Less solar radiation reaches the earth during the more eccentric orbit (outer orbit), a factor that is important in triggering ice ages. (Phil Miller)

Ice Ages—In 19,000- to 23,000-year cycles, the direction in which the earth tilts toward the sun switches. When the Northern Hemisphere is tilted away from the sun during the winter equinox (lower orbit), cold conditions prevail and ice ages take hold. The degree to which the earth tilts on its axis also switches, but in 41,000-year cycles. This is the third factor that contributes to the development of ice ages. (Phil Miller)

(foraminifera and coccoliths) that lived at or near the surface of the ocean. These tiny living things secrete a protective shell of calcium carbonate that is often very ornate, and like terrestrial plants, the coccoliths use photosynthesis to convert water and carbon dioxide into food. When these tiny organisms die, they sink to the seabed, leaving tiny shells that build up into sedimentary deposits on the seafloor. Experts who study these shells, micropaleontologists, can identify different species of foraminifera and coccoliths. In life, each species inhabited a narrow range of sea surface temperatures, and so scientists can analyze the layers to determine if ancient surface waters of the ocean were cooler or warmer than today at the same geographic location.

During the past two decades, analyses of ice cores from Antarctica have provided new information on climate variability during the last 800,000 years. As it falls, snow carries with it atmospheric gases such as carbon dioxide and methane in the same concentration as they appear in the air. Over time, in Antarctica, this snow is compacted from firn to ice, and the record of atmospheric composition is trapped in the bubbles in the ice. A long core of this compressed snow is a record of the earth's climate that stretches back for 800,000 years. These cores show that during a full glacial, the concentration of carbon dioxide averaged 180 parts per million (ppm); during the interglacials, the concentration of this gas averaged 280 ppm. However, human activities since the industrial revolution have been pumping carbon dioxide into the atmosphere in ever greater quantities, and in 2008, the concentration of carbon dioxide in the earth's atmosphere exceeded 380 ppm. As a result, the earth is now warmer than it would be without human activity.

These cores enable us to look back in time and to see how the earth's climate has changed over the eons. If we go back as far as 620,000 years, it seems that there have been seven glacial-interglacial cycles, each of which has lasted between 88,000 and 118,000 years. These cycles are dominated by the cold, glacial phases, as the warmer interglacials have only lasted for between 28,000 and 49,000 years.

As an average human life span is around 75 years, we have little appreciation of cycles that are played out over hundreds of thousands of years—all we can ever see are the aftereffects. The implications for life on earth of these continual oscillations between chilly and warm are huge. Land-living animals can migrate in the face of climatic change, but plants, with their roots fixed firmly in the ground, must simply allow their range to recede and expand with the changing conditions. The glacial periods are not only cold, but also dry, conditions that do not favor the growth of dense forests. During these cold periods, forest cover the world over dwindled, and grassland edged in to replace the trees. As the glacial phase ended, the situation reversed, and the forests moved back into their old range. Animals are free to move around, but specialist forest dwellers dwindled or disappeared altogether, while other animals, more suited to open habitats, thrived. These circumstances opened up new habitats every few thousand years, ideal for the evolution of new species that adapted to fill the new niches. Several of the animals in this book—great beasts like the woolly mammoth, mastodon, and woolly rhinoceros—were cold-adapted species that evolved to take advantage of the habitats created by the glacial-interglacial cycles.

So scientists have worked out that for the last 1.6 million years, earth and its organisms have endured cycles of numbing cold interspersed with warmer periods, but what causes these cycles? After lots of experiments and number crunching over many, many years, scientists now have a good idea of what causes these cycles. Beginning in 1930, Milutin Milankovitch, a Serbian geophysicist and astronomer, spent many long hours, days, and years working out a theory of climate change. He determined that the earth's orbit around the sun is not the simple, consistent, circular route it was always assumed to be. The orbit of the earth around the sun varies from an almost circular path to a very elliptical one (eccentricity) over a roughly 100,000-year cycle. Also, the earth is slightly tilted on its axis of spin, a planetary phenomenon that results in the seasons. This tilt varies between 20.4 degrees and 26.2 degrees over a 41,000-year cycle. One more peculiarity in the way that earth moves through space is that it wobbles on its axis (precession), in the same way as a spinning top set in motion on a flat surface will wobble. One complete cycle takes about 21,000 years, and it also changes the way that observers on the ground see the night sky. Today, the earth's north pole points at the pole star, Polaris, but 13,000 years from now, the pole star will be Vega because of precession. Depending on how these cycles overlap, less solar energy ends up reaching the Northern Hemisphere—the suggested trigger of the cold glacials. More snow falls during these cold spells, less thaws, and the ice sheets start to grow. As they grow, more and more solar energy is reflected back into space by the white snow and ice, and the cooling effect is exacerbated.

Geologically, the current epoch is known as the Holocene. The Holocene is actually a warm interglacial, and it has lasted, so far, for 10,000 years. It's a sobering thought that all recorded human history, at the very beginning of which agricultural civilizations began to replace hunter-gathering as a way of life, has been played out in a relatively warm interglacial. All the known civilizations, all the wars, all the technological advances have come to pass in a narrow, warm window. It would be very naïve for the human race to think that these balmy conditions are going to last forever, and the deep sea and ice cores have shown us that the transition from warm to cold can be astonishingly quick—a couple of decades. These rapid changes are recorded in ice cores from Greenland, and there is no reason to think that the next transition won't be similarly swift. Climate change is a hot topic at the moment; it appears everywhere in the news, and it seems that the earth's climate is beginning to destabilize, which some people think heralds a new climatic age. If this is the case, what are we heading into? The mass media have exhausted the term *global warming*, and it is highly likely that the observed increases in temperature are likely to prolong the present interglacial, the Holocene, but eventually, the natural variations in the earth's orbital elements will lead to another ice age. Over thousands of years, humans, as a species, have become adapted to the relatively easy time afforded by the Holocene. When the earth enters another ice age (and inevitably, it will), our current way of life will be impossible, and the human race will be pushed toward extinction like countless other species over time. In the unlikely event of the earth warming up by a few degrees and staying warm for the next few millennia, the human race would be similarly challenged, and our survival would balance on a knife's edge. In particular, melting of the glacier ice will raise global sea levels, inundating low-lying islands, deltas, and coastal plains. During the last interglacial, sea levels rose by about 6 m, probably due to the near-complete melting of the Greenland ice sheet. Responding to rises in sea level will present major challenges to many nations, including dealing with the displacement of many human populations.

6

MORE THAN 12,500 YEARS AGO

GIANT SHORT-FACED BEAR

Giant Short-Faced Bear—A giant short-faced bear, the largest bear ever, is seen here using its great size to scare a pack of wolves away from their kill—a bison. (Richard Harrington)

Scientific name: *Arctodus simus*
Scientific classification:
   Phylum: Chordata
   Class: Mammalia
   Order: Carnivora
   Family: Ursidae

When did it become extinct? This bear is thought to have died out around 12,500 years ago.

Where did it live? This bear and its close relatives were only found in North America. Their remains have been found from Alaska and the Yukon to Mexico and from the Pacific to the Atlantic coasts.

Thousands of years ago, northern North America was not the land of forest that it is today. Expansive grasslands stretched out toward the horizon, which were populated by great herds of herbivorous mammals, including mammoth, bison, deer, and caribou. Predators like the saber tooth cat, scimitar cat, and dire wolf stalked these herds, and dependent on them were the scavengers. One of these scavengers was the largest bear that has ever lived—a bear so big that even when it was standing on all fours, it could still look a grown man in the face. This was the giant short-faced bear, and in these prehistoric northern climes, it was the dominant carnivorous animal, although it is now widely believed that it was a scavenging animal, rather than an active predator.

This giant bear's closest living relative is the spectacled bear (*Tremarctos ornatus*) of South America, but in appearance, it was unique, with long limbs and a short, wide head. Fully grown, they were enormous—an adult male could have easily tipped the scales at 900 kg (by comparison, a really big male polar bear is around 600 kg). The way the bones of the giant short-faced bear articulate suggest that this huge carnivore was easily able to rear up onto its back legs. A big, standing male was around 3.4 m tall, with a vertical reach extending to around 4.3 m—this is more than 1 m above a basketball hoop. Like modern bears, this extinct brute probably reared up to sniff the air for the telltale odor of meat and to intimidate animals that dared to get between it and its food. We know that the short-faced bear had a big space in its skull for nasal tissue, and its sense of smell was probably very keen—even better than that of modern bears, with their very sensitive noses.

Once this huge bear caught a whiff of some food, it would head for the source. For a long time, it was thought that the favored locomotion of this long-limbed bear was running, but recent research suggests that it moved in the same way as a camel, with what is best described as a pace whereby the two left limbs move together, followed by the right limbs. This is a very efficient gait, and like a speed walker, the bear was able to cover long distances without tiring.

How do we know that this extinct bear was a scavenger? The levels of two types of nitrogen in the bones of an animal (even long-dead ones) can tell us if they were an omnivore or a dedicated carnivore. The nitrogen signature of the short-faced bear's bones suggests that it fed solely on meat, but although it was big, it was not really equipped to be a predator. Its bones seem too slender to have enabled it to tackle the large animals that its big appetite required, and although it was an endurance athlete, it was not fleet of foot enough to catch fast-running prey. In some ways, scavenging is an easy option: you let another animal do the dirty work of killing, the smell of death gets carried on the air, and then you turn up to chase the predators away from their kill with your formidable size. Easy! This is not to say that the short-faced bear didn't actively kill when the opportunity arose, but scavenging seems much more likely. It is easy to imagine the scene of a pack of wolves feasting on the carcass of a young mammoth, only to be scared off by the sight of a giant bear looming over them. With the carcass to itself, the bear could have proceeded to gorge itself on meat. Its teeth and jaws appear to have been sufficiently strong to break the bones of a carcass to get at the nutritious marrow within—the same technique used by the modern-day spotted hyena.

So what became of the giant short-faced bear? How come it can no longer be found lumbering around the northern wilderness, sniffing out carcasses? The long-standing belief was that this giant was outcompeted by the brown bear as the latter species migrated into North America via the Bering land bridge. As it is now assumed that the giant short-faced bear was a scavenger, the two species only came into direct competition in certain circumstances, for example, in the event of dwindling resources. The brown bear is an omnivore that gets its calories from a wide variety of sources, of which carrion makes up only a fraction. Competition may have played a part in the demise of this giant; climate change was probably the most important factor. Toward the end of the last glaciation, the increase in global temperatures was responsible for the disappearance of northern grasslands, as the warmer, wetter conditions favored the growth of forests. These boreal forests cover vast swathes of the Northern Hemisphere today, and thousands of years ago, they probably deprived the giant bears of prime scavenging territory. The dwindling populations under pressure due to habitat loss, competition, and even disease transmitted by the spreading brown bears may have been sufficient to drive the giant short-faced bear to extinction.

+ The giant short-faced bear is known to have existed for at least 800,000 years, and possibly far longer. In that time, the species experienced many global warming and cooling events, lending support to the theory that it was not one single factor that led to the extinction of this species.
+ The remains of this bear have been found in caves. The bones discovered in Potter Creek Cave, California, are all from females, indicating that this species may have made dens in such places to give birth and to raise their young, until they were big enough to face the rigors of the outside world.
+ The bones of this bear even provide a window into some of the diseases from which they suffered. There is evidence of osteomyelitis, tuberculosis-like diseases, and syphilis-like infections.
+ Humans definitely hunted the large animals on which the giant short-faced bear was dependent, and this may have been another factor contributing to the bear's extinction.

Further Reading: Matheus, P. E. "Diet and Co-ecology of Pleistocene Short-Faced and Brown Bears in Eastern Beringia." *Quaternary Research* 44 (1995): 447–53; Voorhies, M. R., and R. G. Corner. "Ice Age Superpredators." *University of Nebraska State Museum, Museum Notes* 70 (1982): 1–4.

FLORES HUMAN

Scientific name: *Homo floresiensis*
Scientific classification:
  Phylum: Chordata
  Class: Mammalia
  Order: Primates
  Family: Hominidae

Flores Human—Here a female *Homo floresiensis*, barely 1 m tall, walks back to her group's cave with a large rodent she has killed. (Phil Miller)

When did it become extinct? The most recent remains of this hominid are 18,000 years
old, but it is very possible that it survived well into historic times.

Where did it live? The bones of Flores man have only been found on the island of Flores,
Indonesia.

In 2004, a group of scientists revealed to the world what they found in a cave on the
island of Flores, Indonesia. The story featured in the news all over the world, and their
discovery could be one of the most important paleoanthropological discoveries ever made.
Almost 6 m beneath the floor of a large limestone cave called "Liang Bua," the team of Australian and Indonesian scientists found a partial skeleton of a human, but one that was quite
unlike anything that had ever been seen before. Although the skeleton was not complete,
there was enough to see that it was an adult female (she was probably around 30 years old
when she died), and the most astonishing thing about the find was the size of the individual.
Fully grown, she was no taller than a three-year-old child—about 1 m tall, with a brain no
bigger than a chimpanzee's.

Ever since the scientists published their discovery in the journal *Nature*, there has been
heated debate on exactly what the skeleton represents. Is it a pygmy modern human, a modern human with a disease or anatomical abnormality, or a genuinely new species? Current
opinion swings in favor of the skeleton being of a new species of human that may have
evolved in isolation on the island of Flores from a *Homo erectus*–like ancestor. Another
amazing thing about the skeleton was its age. The bones were not fossilized, nor were they
covered in calcium carbonate. They were actually very delicate, with the consistency of wet
blotting paper. The material around the bones was aged using modern techniques, and it
turned out that they were around 18,000 years old. Before this discovery, it was thought
that the Neanderthals, the last species of human other than our own species, died out
around 28,000 years ago. If the Flores discovery is a genuinely new species, modern humans
had shared the earth with another species of human, albeit a miniature one, up until at least
18,000 years ago, which, in geological terms, is the blink of an eye.

Why was this human so tiny? The diminutive size of the Flores human could be due to a
phenomenon known as the island rule. This phenomenon can be seen on islands all over the
world. It seems, that in some cases, any animal larger than a rabbit that finds itself marooned
on an island shrinks, but for some animals smaller than a rabbit, the reverse is true, and they
develop into giants. Survival on an island can be tough; food may be in short supply, and dis-
persing to new habitats is not an option. Therefore, if you are a big animal, it makes sense to
shrink as a smaller body requires less energy than a big body. Scholars always assumed that
humans were beyond this general biological rule because they can make fire to keep warm and
use a host of other ways to cheat the environment. Perhaps the ancestors of the Flores humans
were less adaptable than modern humans, and the conditions favored a smaller body size.

Along with the bones, a great number of stone artifacts were also found. Many of these
are simple stone tools, but some are much more sophisticated and seem to be designed for
specific purposes. Again, debate rages over whether these tools were made by Flores humans
or modern humans who occupied the cave at a later date Their size suggests that they were
wielded by small hands, but until more bones and tools are unearthed, it will be difficult
to know for sure. Regardless of the tools Flores humans fashioned, they hunted the Flores

animals for food. Many of the animal bones found along with the Flores human skeleton belong to an animal called a stegodon, a small-bodied distant relative of modern elephants, which had also gone through a shrinking process, until it was a dwarf compared to its close relatives on the mainland. Some of the stegodon bones bear the marks of butchery and burning. Is this one of the animals these diminutive humans hunted? A fully grown stegodon is small by modern elephant standards, but it still weighed in the region of 1,000 kg, and a lone, 25-kg human could never have brought down one of these animals; therefore the Flores humans must have hunted in teams, coordinating their efforts to subdue their large quarry.

If the Flores humans were able to hunt cooperatively, use fire, and make and use tools, they must have been intelligent, yet they had a tiny brain, about one-third the size of ours. Before the discovery of the Flores human, the accepted theory was that brain size and intelligence in hominids went hand in hand (the bigger the brain, the more intelligent the hominid). The bones unearthed in Liang Bua cave have challenged this long-held belief. Perhaps brain size is not the last word when it comes to intelligence; perhaps the most crucial factor is the way in which all the cells in the brain are linked together. This is but one of the many contentions surrounding the discovery and study of this fascinating skeleton.

The scientists who made this initial discovery plan to return to the site to make more excavations. If they find more miniature skeletons, or even just skulls, it will prove beyond any reasonable doubt that Flores was once home to a species of tiny human. If this is correct, then what happened to these diminutive humans? Around 12,000 years ago, an immense volcanic eruption shook the area, and it is possible that this caused the demise of this species. However, Flores lore tells of mysterious dwarves called *ebu gogo* (literally translated, this means "grandmother who eats anything"). According to folklore, the *ebu gogo* were alive when Portuguese trading ships reached Flores 400 years ago, and some islanders believe that they were still around up until 100 years ago. Whether these accounts are genuinely a folk memory of extinct Flores humans or simply fireside stories will never be known, but they are nonetheless very interesting.

+ In total, Liang Bua cave yielded bones from eight individuals of the Flores human, but so far, only one cranium has been discovered. More excavations on the island will hopefully reveal a complete skeleton of this hominid.

+ The origins of the Flores human are unclear. Tools aged at 840,000 years old, thought to be the work of *Homo erectus*, have also been found on the island. The skull of the Flores human has many similarities with the known *Homo erectus* skulls, and as *Homo erectus* is the only hominid that we know for sure reached the Far East (apart from our own species), we can be reasonably confident that the Flores human descended from a population of *Homo erectus* that somehow became marooned on this Indonesian island.

+ The scientists who discovered the Flores human have speculated that other Indonesian islands may also have had their own unique populations of human, the remains of which are still waiting to be discovered.

+ Sightings of a short, bipedal hominid covered in short fur have been reported for at least 100 years from the island of Sumatra. Known by the islanders as the *orang pendek*, this animal is said to be around 150 cm tall and to reside in the dense rainforest. Is it possible that another species of hominid has escaped detection by the scientific world and is living in the rapidly dwindling forests of this huge Indonesian island?

Further Reading: Morwood, M.J., R.P. Soejono, R.G. Roberts, T. Sutikna, C.S.M. Turney, K.E. Westaway, W.J. Rink, X. Zhao, G.D. van den Bergh, D. Rokus Awe, D.R. Hobbs, M.W. Moore, M.I. Bird, and L.K. Fifield. "Archaeology and Age of a New Hominin from Flores in Eastern Indonesia." *Nature* 431 (2004): 1087–91; Wong, K. "The Littlest Human: A Spectacular Find in Indonesia Reveals That a Strikingly Different Hominid Shared the Earth with Our Kind in the Not So Distant Past." *Scientific American*, February 2005; Brown, P., T. Sutikna, M.J. Morwood, R.P. Soejono, E. Jatmiko, E. Wahyu Saptomo, and D. Rokus Awe. "A New Small-Bodied Hominin from the Late Pleistocene of Flores, Indonesia." *Nature* 431 (2004): 1055–61; Morwood, M.J., P. Brown, E. Jatmiko, T. Sutikna, E. Wahyu Saptomo, K.E. Westaway, D. Rokus Awe, R.G. Roberts, T. Maeda, S. Wasisto, and T. Djubiantono. "Further Evidence for Small-Bodied Homininss from the Late Pleistocene of Flores, Indonesia." *Nature* 437 (2005): 1012–17; Obendorf, P.J., C.E. Oxnard, and B.J. Kefford. "Are the Small Human-like Fossils Found on Flores Human Endemic Cretins?" *Proceedings of the Royal Society B: Biological Sciences* (2008), doi:10.1098/rspb.2007.1488.

GIANT BISON

Giant Bison—This skeleton measures just over 2 m from the floor to the top of the tallest vertebral spine. In life, the animal would have been closer to 2.5 m tall, with horns at least 2 m across. (Royal Saskatchewan Museum)

Scientific name: *Bison latifrons*
Scientific classification:
    Phylum: Chordata
    Class: Mammalia
    Order: Artiodactyla
    Family: Bovidae
When did it become extinct? The giant bison became extinct sometime between 21,000
    and 30,000 years ago.
Where did it live? This bison ranged widely across what are now the United States and
    southern Canada.

The modern American bison (*Bison bison*) is the quintessential American mammal. It is
thought that 60 to 100 million bison roamed North America before the arrival of Europeans.
As the settlers moved westward, they ravaged the bison herds until the species teetered on
the brink of extinction. Fortunately, the bison received protection, and today, there are strong
populations of this animal in several national parks in the United States and Canada as well
as those living on private ranches.

The bison we know today is one of the last vestiges of the American megafauna, and
these lands were actually home to several different kinds of bison, all of which originated
from animals that migrated into North America from Asia via the Bering land bridge.
It is still not clear if these fossils represent distinct species or geographical and temporal
variants of a single, highly variable bison species. The ancestors of the bison evolved in
Eurasia around 2 to 3 million years ago, and from there they spread, eventually reaching
North America around 300,000 years ago. The North American continent was a land of
opportunity, and these ancestral bison diversified into a range of forms, the most impressive of which was the giant bison. The modern plains bison is a big animal, with males
reaching 2 m at the shoulder and 900 kg in weight; however, they would be dwarfed by a
giant bison. This extinct species was around 2.5 m at the shoulder and could have weighed
as much as 1,800 kg. Not only were they big, but the giant bison also had incredible horns.
Like all bovids, the giant bison's horns were composed of a bone core surrounded by a
keratin sheath. The sheath rots away to nothing after being buried for thousands of years,
leaving us with just the bony cores curving out from the big skull. Some of these skulls have
a horn span of just over 2 m, but in life, the keratin sheath made the span even wider, as is
shown by a Californian specimen in which the outer sheaths were replaced by a sediment
cast. Today's male American bison are far larger than the females, but this sexual dimorphism was even more pronounced in the giant bison. A fully grown male giant bison with
its huge, shaggy forequarters and amazing horns must have stood out like a beacon amid
the much smaller females.

The living bison is divided into two subspecies: the plains bison (*B. bison bison*) and the
wood bison (*B. bison athabascae*). It is thought that the latter species has more in common
with the giant bison in terms of behavior. The giant bison is not thought to have lived in
the immense herds that the plains bison forms because despite its size, relatively few fossil
specimens have been found in comparison to later bison. It may have formed, instead, small,
close-knit family groups. The fossils of the giant bison have been found over a wide geographic area, and this could indicate that the animal was able to live in a variety of habitats,

including forests and parklands as well as steppe grasslands, where it grazed on and browsed a wide range of plants.

Exactly how the giant bison used its enormous horns is not clear, but they were definitely important when it came to the breeding season. Males must have fought for the right to mate with as many females as possible, but it is likely that the males with the most impressive horns averted disputes through display, by simply intimidating their rivals with their size. Pleistocene North America supported a diverse population of predatory mammals, many of which were a match for a bison, even a giant one. The larger saber tooth cats, American lions, and wolves hunting in packs may have been able to overpower a fully grown giant bison, but tackling an adult male with its vicious horns and great strength must have been very dangerous. The predators of the giant bison most likely focused their attention on calves and on old and sick adults.

The giant bison seems to have vanished before humans arrived in North America, but it is unlikely it became extinct in the normal sense. As the giant bison adapted to the ever changing American landscape, it evolved into the smaller fossil species, the ancient bison (*Bison antiquus*). *Bison antiquus* lived between about 20,000 and 10,000 years ago and, in turn, evolved into the modern bison. Mitochondrial DNA recovered from *Bison antiquus* is very similar to that of modern bison, demonstrating the association. No DNA has yet been recovered from the fossils of the giant bison, but there is a clear reduction in size moving from the giant bison, to the ancient bison, to the modern bison, providing a good example of evolutionary change.

The first humans to colonize North America, the Clovis culture, known from their widespread, distinctive flint arrowheads and spearheads, undoubtedly knew the ancient bison—the descendents of the giant bison—which, by Clovis times (about 13,000 years ago), was a grassland animal swelling in numbers. Along with the mammoths and the mastodons, this was still one of the larger land mammals of North America, and killing an adult probably provided a small group of humans with enough food for many weeks and an abundance of raw materials for making tools, shelter, and clothing. Is it possible that human hunting caused the demise of this bison? The likely answer is no. As with all the other great beasts that once roamed North America, we cannot attribute the disappearance of the ancient bison to a single event or factor. For almost the last 2 million years, the earth's flora and fauna have had to adapt to massive, cyclic climatic changes, some of which have been very abrupt: the glaciations and their associated interglacials (see the "Extinction Insight" in chapter 5). The inhabitants of the high and low latitudes have been most affected by these changes, but animals have the ability to migrate in the face of worsening conditions, even if it means that their populations may shrink. When humans arrived in North America, the giant bison had also felt the squeeze of climate change and had evolved into a smaller form, which in turn evolved into the smaller modern bison. Hunting probably had a considerable impact on populations, but bison were distinctive in being able to withstand these pressures and even to increase in number, until the arrival of the gun finally drove them to near-extinction in the nineteenth century.

+ It is thought that the bison that migrated into North America from Asia were steppe bison (*Bison priscus*). They were, in turn, ancestral to the giant bison, which, through evolutionary change, spawned the two American bison subspecies we know today.
+ In the United States and Canada, archeologists have unearthed what appear to be kill sites: locations where the first Americans processed the bodies of ancient bison for their meat, skin, bone, and sinew. Some of these sites have yielded the remains of

hundreds of bison, which goes to show how important these animals were to the survival of prehistoric humans in North America.

+ A species of bison also survives in Europe. The European bison, or wisent (*Bison bonasus*), is a forest-dwelling animal that once ranged over much of Eurasia. Hunting depleted its numbers, until the last wild specimen was killed in 1927. Fortunately, several wisent were kept in zoos and private collections, and these were used to start a reintroduction program. Today, thanks to reintroduction and protection, the largest European land animal can be found in several eastern European countries.

Further Reading: Guthrie, R.D. "Bison and Man in North America." *Canadian Journal of Anthropology* 1 (1980): 55–73; McDonald, J.N. *North American Bison, Their Classification and Evolution.* Berkeley: University of California Press, 1981.

HOMO ERECTUS

Scientific name: *Homo erectus*
Scientific classification:
   Phylum: Chordata
   Class: Mammalia
   Order: Primates
   Family: Hominidae
When did it become extinct? Exactly when *Homo erectus* died out has divided scientists for years. Some paleontologists believe that isolated populations of this hominid may have survived in Southeast Asia until fewer than 100,000 years ago.
Where did it live? Fossils of *Homo erectus* have been found in Africa, the Republic of Georgia, China, and Indonesia.

Eugène Dubois, a Dutch anatomist, set off for the Far East in the 1880s, intent on finding fossils of the missing link between apes and humans. He searched fruitlessly in New Zealand before shifting his attention to Java, one of the large Indonesian islands. Amazingly, and to the disbelief of the scientific community, his Javan expedition was a success, as he found the skullcap and femur of one of our ancient ancestors. Whether Dubois's find was due to excellent judgment and insight or plain luck is a source of some academic debate. Given that the bones of our very ancient ancestors are extremely rare, this find is actually more remarkable than finding a needle in a haystack (at least with a haystack, you can use a metal detector!). It later turned out that these bones were not from the so-called missing link, but Dubois's discovery was nonetheless a major breakthrough in the area of research that attempts to understand our origins.

The owner of the bones Dubois discovered was named *Homo erectus* (erect man), and up until 1984, all the known remains of this extinct hominid species would have fitted quite comfortably in a large shoe box—such was their rarity. This all changed with the discovery of an almost complete skeleton in East Africa that has become known as Turkana Boy. Turkana Boy gave the world its first glimpse of what an almost entire *Homo erectus* skeleton looked like, and it became clear that they were the first of our ancient ancestors to have a truly human look, with a very erect posture and long legs. The pelvis of *Homo erectus* is

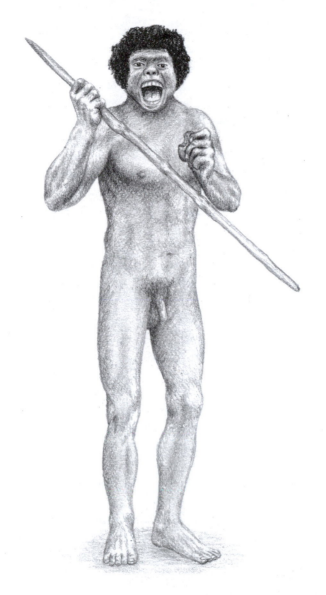

Homo erectus—*Homo erectus* was a strong athlete and the first of our ancient ancestors to disperse widely from Africa, reaching at least as far as Indonesia. (Phil Miller)

narrow compared to a modern human, a feature that made them very accomplished runners. Apart from the skull, their skeleton is very similar to our own, and it would take an expert to tell them apart. Adult males were around 1.8 m tall and physically very strong. The skull of *Homo erectus* it what really sets this species apart from us. First, the brow ridges of the skull were very pronounced, and it also lacked a chin, but most important of all is the cranium and what it contained. The cranium of *Homo erectus* was smaller than our own and carried a brain that was only around 75 percent of the size of an average modern human's. Because the frontal lobes of *Homo erectus*'s brain were very small compared to our own, its forehead was very sloping and shallow. As with other human ancestors, the lower jaw of

Homo erectus was robust and equipped with big teeth. This mandible was powered by large muscles and was undoubtedly suited to chewing tough food.

Using a complete skull of *Homo erectus*, anatomists and artists can build up a picture of the face of this extinct hominid. If you stare into the face of one of these reconstructions, you can see yourself, but the overall impression is of an animal that was barely human. The mental capabilities of *Homo erectus* can only be guessed. A frequent question is whether these hominids were able to express themselves with language, and detailed studies of Turkana Boy suggest that their power of speech was very minimal, perhaps limited to simple sounds—the precursors of complex speech. We do know that they made tools, as stone artifacts have been found at the same locations as their bones and from other locations around the world. The bones of *Homo erectus* are so rare that these tools give us a better picture of just how geographically widespread this hominid was.

In a short period of geologic time, this hominid dispersed from Africa to Eurasia in the north and China and Indonesia in the east (and possibly even farther). These movements suggest that *Homo erectus* was capable of solving complex practical problems as they were confronted by treacherous bodies of water and other seemingly insurmountable barriers. With narrow hips and long legs, *Homo erectus* was a natural athlete, and this may have been crucially important in allowing them to disperse far and wide from where they first evolved.

There is also some tantalizing evidence that *Homo erectus* harnessed and used fire, one of the major technological leaps in human evolution. Fire made food safer and more palatable and kept predators at bay as well as having a multitude of other uses. *Homo erectus* stone implements may be just a fraction of what these hominids were capable of creating. They could have produced a range of different tools using plants and various bits of animal, but if these have stood the tests of time anywhere and not rotted away completely, they have not yet come to light.

Homo erectus was undoubtedly a physically strong hominid, but was it an active predator or a scavenger on the kills made by predators such as big cats? Hunting requires a lot of time and energy, and it can also be very dangerous. Scavenging is less dangerous, but it is not easy, especially if you are planning on stealing a carcass from beneath the nose of a saber tooth cat. However, the risks of scavenging are outweighed by the rewards of a huge amount of fresh meat.

The oldest *Homo erectus* fossils are around 1.8 million years old, and the most recent remains could be fewer than 100,000 years old, so this was a very successful and widespread species. What happened to *Homo erectus*? The likely cause of the extinction of *Homo erectus* was competition with modern humans, who treaded the same paths out of Africa, eventually colonizing almost the entire globe. Our species, *Homo sapiens*, was probably inferior to *Homo erectus* in terms of brute strength and stamina, but our unparalleled advantage was our brain and the language and ingenuity it gives us.

+ As *Homo erectus* evolved on the hot, arid plains of equatorial Africa, it was adapted to cope with the powerful sun's rays. An upright stance presents less of the body's surface to the heat of the sun, and it was probably hairless, which allows the evaporation of sweat to cool the underlying blood. Its skin was darkened with melanin, a pigment that protects the skin cells from the damaging effects of the sun's ultraviolet radiation.

+ Exactly how *Homo erectus* crossed from the mainland and reached many of the Indonesian islands is still a mystery. Low sea levels could have revealed land bridges, but there is also the possibility that *Homo erectus* was the earliest seafarer. This hominid species may have had sufficient mental ability to figure out a way of crossing open stretches of water to reach the island of Flores well over 800,000 years ago.

+ In modern humans, the average difference in size between males and females is quite small, but adult *Homo erectus* males were 20 to 30 percent bigger than adult females

Further Reading: Brown, F., J. Harris, R. Leakey, and A. Walker. "Early Homo Erectus Skeleton from West Lake Turkana, Kenya." *Nature* 316 (1985): 788–92; Swisher, C.C. "Dating Hominid Sites in Indonesia." *Science* 266 (1994): 1727; Rukang, W., and L. Shenglong. "Peking Man." *Scientific American* 248 (1983): 86–94.

NEANDERTHAL

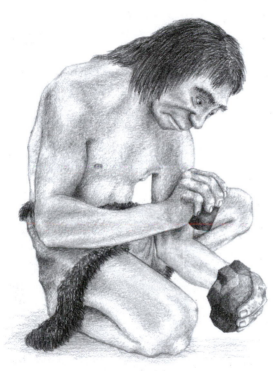

Neanderthal—The Neanderthals were the first Europeans. They had large brains and were powerfully built, yet they died out. Exactly what happened to them is one of the greatest mysteries in human evolution. (Phil Miller)

Scientific name: *Homo neanderthalensis*

Scientific classification:
Phylum: Chordata
Class: Mammalia
Order: Primates
Family: Hominidae

When did it become extinct? The most recent remains of Neanderthals have been dated at around 28,000 years old, and it is unlikely that they survived into more recent times.

Where did it live? Neanderthals lived throughout Europe, into the Middle East and southern Siberia.

For a long time, the word *Neanderthal* was synonymous with lumbering cavemen, and following the first official discovery of a partial skeleton of a human in Germany, Victorian scientists had a field day in portraying this extinct human as a stooped, troll-like beast. True, the Neanderthals may have had quite a brutish appearance by our terms, what with their stout limbs, broad chests, projecting brows, and powerful jaws, but in recent times, our view of our long-dead relatives has changed, as more remains have come to light. These remains are not ridiculously common, but they do tell us a story of a species—another human species—with which we once shared the earth, and it has become clear that Neanderthals, far from being the knuckle-dragging ogres of Victorian imagination, were actually a sophisticated and successful species.

We know from artifacts that have come to light that the Neanderthals made tools, and their ability in this regard was not far behind that of the Cro-Magnons (modern humans— our species—in Europe) who replaced them. Ancient unearthed tools thought to have been fashioned by Neanderthal hands provide us with an intriguing yet incomplete picture of how our relatives lived. How did they go about catching their food, for instance? Their teeth and jaws are typically those of a vegetarian-omnivore, but analysis of their bone chemistry has led some people to speculate that their diet was mainly meat, and if this came from living animals, how did they catch and subdue their prey? For a long time, scientists believed that the Neanderthals were only capable of wrestling with their prey and hacking it to death with stone hand axes or similar tools. However, recent finds paint a picture of a human that could fashion spears and other weapons to strike at prey from a distance. With this said, they probably had to close in for the killer blow, using their great strength to finish off the prey. The bones of Neanderthals that have been discovered over the past 150 years or so often show signs of injury, such as bone fractures and breaks, that may have been inflicted when these extinct humans were tackling and killing wild beasts.

Even more surprising is the fact that many of these bone breaks and fractures were healed, an observation that gives us a tantalizing glimpse of how these extinct humans interacted with one another. Injured Neanderthals must have been cared for by those around them, perhaps in a family group or even a tribe, because a solitary Neanderthal with a broken leg would not have survived long enough for the broken bones to heal. We can assume that Neanderthals cared for their sick, and perhaps even their elderly, as some bones are from individuals more than 50 years old, which was a grand old age many thousands of years ago. What other characteristics did they share with us? Did they have language? It is thought by some experts that Neanderthals could speak, as a hyoid bone—a small bone that is part of the speech apparatus—was found with a Neanderthal skeleton in Kebara Cave, Israel, in the 1980s, and what is preserved is similar to ours. However, this bone only gives us an idea of what sounds the Neanderthal could make as the bone works with the soft tissues of the larynx to produce the sounds we know as words. Without these soft tissues, it is impossible to know exactly what sounds the Neanderthals were capable of making, but it has been suggested that Neanderthal language was not as elaborate as our own.

Along with some form of vocal communication, the Neanderthals buried their dead. Some paleontologists have suggested that the Neanderthals adorned the bodies of their dead with flowers, but this theory is very controversial and is based on the discovery of one skeleton commonly known as the Shanidar burial. If it were true, such a ritual would indicate that these long-dead humans had a complex culture that possibly included religion and a concept of life after death.

So what happened to the Neanderthals? This is a big mystery, but numerous theories attempt to explain the disappearance of this other species of human. A popular one is that our ancestors, on their migration north from Africa, moved into the lands of the Neanderthals and eventually outcompeted them, even possibly going out of their way to eradicate them. A second popular theory is that modern humans and Neanderthals interbred to such an extent that the characteristics of Neanderthals were diluted so much that we cannot see them today. If this theory is correct, then modern humans, especially those of us with roots in northern and western Europe, still carry Neanderthal genes. It has even been suggested that ginger hair is a Neanderthal trait that has survived into the modern day, although this trait probably evolved independently in modern humans and Neanderthals.

These theories aside, we know that the Neanderthal world went through some major shifts as the ice sheets advanced and then retreated and the flora and fauna of the Neanderthal lands were massively influenced by these changes. Perhaps the Neanderthals succumbed to the combination of the relentless spread of our ancestors and a changing landscape and climate, but maybe, just maybe, the Neanderthals live on in us.

- Neanderthals evolved in Europe, and it is thought that their ancestors, an earlier form of *Homo erectus*, left Africa and dispersed over much of what we know as the Old World today.
- Neanderthals were often portrayed as an unsuccessful species that eventually succumbed to the more sophisticated Cro-Magnons, but in actual fact, this extinct species of human survived for at least 250,000 years and was well adapted to a very harsh environment. In comparison, our own species has only been around for a mere 120,000 years.
- Our knowledge of what prehistoric humans were capable of making is limited to objects made from material that can survive the ravages of time, for example, stone and bone. Much of the wear on Neanderthal stone tools comes from wood working, yet we have no idea what they were whittling as it has all rotted away, except for one solitary bowl (discovered in Abric Romani, Spain) and spears (from Schoeningen and Lehringen, Germany).
- A Neanderthal's brain was actually as large as ours, but the skull was a very different shape. The high forehead of a modern human skull accommodates the well-developed frontal lobes, which may be the seat of the higher mental processes that characterize modern humans. There is little in the artifactual record of Neanderthal behavior to suggest that they possessed symbolic thought, as we do.

Further Reading: Speth, J. D., and E. Tchemov. "The Role of Hunting and Scavenging in Neandertal Procurement Strategies." In *Neandertals and Modern Humans in Western Asia*, edited by T. Akazawa, K. Aoki, and O. Bar-Yosef, 223–29. New York: Plenum Press, 1998; Thieme, H. "Lower Paleolithic Hunting Spears from Germany." *Nature* 385 (1997): 807–10; Boëda, E., J. M. Geneste, C. Griggo, N. Mercier, S. Muhesen, J. L. Reyss, A. Taha, and H. Valladas. "A Levallois Point Embedded in the Vertebra of a Wild Ass (*Equus africanus*): Hafting, Projectiles and Mousterian Hunting Weapons." *Antiquity* 73 (1999): 394–402.

MARSUPIAL LION

Scientific name: *Thylacoleo carnifex*
Scientific classification:
   Phylum: Chordata
   Class: Mammalia
   Order: Diprotodontia
   Family: Thylacoleonidae

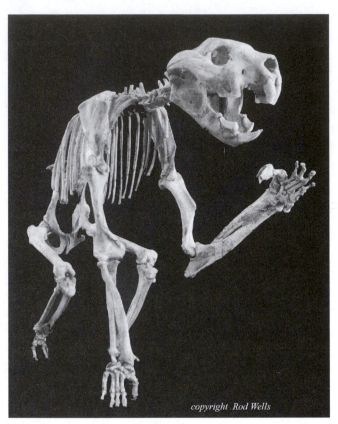

copyright Rod Wells

Marsupial Lion—The marsupial lion is an Australian oddity. Its grasping thumb and fearsome teeth are clearly visible in this image. (Rod Wells)

When did it become extinct? This marsupial appears to have gone extinct approximately 40,000 years ago.

Where did it live? The marsupial lion was found only in Australia.

The word *marsupial* conjures up images of cuddly creatures like the kangaroo, koala bear, and wombat, but many thousands of years ago, some very different marsupials stalked Australia, and one of these, the marsupial lion—a relative of the wombats and kangaroos—was probably the most bizarre pouched mammal that has ever lived. Sir Richard Owen, the renowned Victorian paleontologist, was first to describe this animal from a small collection of skull fragments, and in a 1859 Royal Society paper he said these bones must have come from "the fellest and most destructive of predatory beasts." He described it as a marsupial "lion." For many years, the deductions of Richard Owen were questioned as this was an extinct marsupial whose closest relatives were vegetarians. However, over time, more remains of this animal came to light, and in 1966, the first almost complete but heavily calcium carbonate–encrusted skeleton was discovered. This reawakened the debate about the feeding habits of this strange animal. The specimen proved difficult to prepare, but then, in 1969, better preserved specimens were discovered in the Naracoorte Caves, which proved beyond any reasonable doubt that Owen's long-extinct marsupial must have been a meat eater, not simply a scavenger either, but, very probably a well-adapted predator.

Since the 1960s, more skeletal remains have come to light, largely from cave deposits, and the marsupial lion has secured its place as one of the most remarkable mammals that has ever lived. In terms of size, the marsupial lion was about the same size as a modern lioness. They were around 75 cm at the shoulder and 150 cm long, and it has been estimated that the heaviest individuals were around 160 kg. Scientists can tell a great deal about where an animal spent its time by looking at its bones, and although it has been suggested that this marsupial was an animal capable of climbing trees, it is now believed that the marsupial lion skulked around on the ground, where it ambushed its prey and perhaps dragged it into caves or up into trees, as leopards do. Not only can we tell where an extinct animal lived, but we can also get a good idea of how it moved by looking at the proportions of the limbs, and it seems that this marsupial lion was no long-distance athlete; instead, it probably ambled about, employing short bursts of speed when the need arose.

The most amazing thing about this animal's skeleton is the skull—it's big and heavy, with some incredible teeth. This marsupial had the most specialized dentition of any carnivorous mammal. Carnivorous placental mammals have enlarged canine teeth for stabbing their prey, but the marsupial lion's canines are small and probably close to useless. The two pairs of incisors, on the other hand, are big and pointy, giving the skull an appearance that is reminiscent of a large rodent. Further down the mouth are enormous premolars that can be as much as 60 mm long. These incredible cheek teeth must have worked like a pair of bolt cutters—slicing through the flesh of prey—powered by the big jaw muscles. The marsupial lion's other intriguing weapon was the clawed thumb on each of its forepaws. Although this digit wasn't a true opposable thumb like ours, it could still be used to exert a very powerful grip, driving the sharp, retractable claw into whatever unfortunate victim the marsupial lion had captured. The presence of this thumb is another reason for the belief that the marsupial lion was a tree-dwelling creature as it would have enabled a good grip on branches and tree trunks; however, it is now widely believed that the thumb was primarily for grabbing and subduing prey. Once the prey was immobilized, the fearsome teeth could be brought into action to deliver the killer bite. The pointy incisors were probably used to break the neck and sever the spinal cord, before the heavy-duty premolars were used to bite chunks of flesh from the dead body of the prey.

Bones can provide us with a sketch of how an animal lived, but for the fine detail, we must resort to deduction. For example, we can never know for sure what animals the marsupial lion preyed on or how it hunted them, but its size and teeth lead us to the conclusion that it must have killed and eaten fairly large animals. In the same cave deposits that have yielded the remains of the marsupial lion, paleontologists have found the hind leg bones of kangaroos and wombats bearing large, opposing, V-shaped cuts that perfectly match the cheek teeth of the marsupial lion, suggesting that they were the victims of this predator. We can be fairly certain that the marsupial lion was a specialist predator because it possessed so many unique features that bear no resemblance to any other predatory marsupials we know, alive or dead. We do know that the marsupial lion shared its home with that other great antipodean predator, the thylacine, and for two top predators to have coexisted in space and time, they must have lived in quite different ways. It's plausible to think of the thylacine as a wolflike predator, using its stamina to chase down prey, and the marsupial lion as more of an ambusher, taking its prey unaware and dispatching it with its battery of weapons.

Along with many of the other unusual mammals that once roamed Australia, the marsupial lion became extinct around 40,000 years ago. This date coincides with a period of increasing aridity in Australia, which reached a climax some 20,000 years ago and an increase in the abundance of charcoal in the fossil record, suggesting a change in the frequency of fires. We also know that humans reached Australia around 60,000 years ago. Humans undoubtedly hunted the prey of the marsupial lion, and it could have been a combination of competition with humans and a fire-induced vegetation change brought about by humans as well as climate change that forced these remarkable Australian mammals into extinction.

+ The mammals are divided into three groups: the monotremes (duck-billed platypus and echidnas), marsupials, and placentals. The latter have become the most widespread and successful of all the mammals, while the marsupials are at their most diverse in Australia and South America. As marsupial females give birth to an embryo that spends the rest of its early development locked onto a teat in a pouch (marsupium), the evolution of wholly aquatic forms was impossible.

+ Some recent scientific research showed that, pound for pound, the marsupial lion had one of the strongest bites of any predatory, land-living mammal. The bite of a 100-kg marsupial lion was at least as powerful as that of a 250-kg lion.

+ The Nullarbor Plain in Australia is riddled with cave systems, some of which are connected to the surface by sinkholes. In 2002, a group of cavers exploring one of these tunnels found a chamber containing the bones of numerous, long-dead beasts spanning a period of time from 195,000 to 790,000 years ago (see the "Extinction Insight" in this chapter). These animals had been roaming around on the plains and had tumbled to their deaths through the cave entrance. The cavers' torches illuminated the finest marsupial lion skeletons that have ever been found—lying on the cave floor in the same position in which they had died thousands of years previously.

Further Reading: Wroe, S., C. McHenry, and J. Thomason. "Bite Club: Comparative Bite Force in Big Biting Mammals and the Prediction of Predatory Behaviour in Fossil Taxa." *Proceedings of the Royal Society B: Biological Sciences* 272 (2005): 619–25; Prideaux, G.J., J.A. Long, L.K. Ayliffe, J.C. Hellstrom, B. Pillans, W.E. Boles, M.N. Hutchinson, R.G. Roberts, M.L. Cupper, L.J. Arnolds, P.D. Devine, and N.M. Warburton. "An Arid-Adapted Middle Pleistocene Vertebrate Fauna from South-Central Australia." *Nature* 445 (2007): 422–25.

DIPROTODON

Scientific name: *Diprotodon* sp.
Scientific classification:
 Phylum: Chordata
 Class: Mammalia
 Order: Diprotodontia
 Family: Diprotodontidae
When did it become extinct? The most recent remains of a diprotodon are the 30,000-year-old bones from Cuddie Springs in southeastern Australia.
Where did it live? The diprotodons were found only in Australia.

Like all of the landmasses on earth, Australia was once home to an array of large animals known as megafauna. The Australian assemblage of giant beasts included massive

Diprotodon—This reconstruction of a large diprotodon doesn't really convey its rhinoceros-like proportions. (Australian Museum)

marsupials, big flightless birds, and monster reptiles. Today, these are all gone, and the largest living marsupial is the red kangaroo (*Macropus rufus*). A fully grown male stands around 1.8 m tall and weighs in at about 90 kg. Thousands of years ago, the red kangaroo was even larger than it is today, but it was still dwarfed by the largest of the diprotodons, which were up to 1.8 m at the shoulder, 4 m in length, and 3 tonnes in weight—the only land animals alive today that are larger are the elephant, hippopotamus, and two species of rhinoceros. These giant marsupials looked a lot like big wombats, and the living wombats and koalas are actually the closest living relatives of these extinct beasts. There were several species of diprotodons—experts disagree on the exact number—but they ranged in size from 500-kg, bear-sized creatures to the aforementioned giants.

For hundreds of thousands of years, these giant marsupials were very widespread, as their bones have been found all over Australia. In some places, such as Lake Callabonna in South Australia, lots of diprotodon skeletons have been found together. Hair thought to be from a diprotodon has also been found as well as footprints preserved in the hardened surfaces of old lake beds. These impressions show that the diprotodon had hairy feet, and we can assume that the whole animal was covered with fur and was not naked like a rhinoceros. These footprints, the places in which they have been found, and the delicate skeletal structure of the diprotodon's feet suggest that these were animals that spent a lot of their time padding around on the soft earth and mud bordering lakes and rivers.

The diprotodon had a big skull, and like its feet, this was also quite fragile, with lots of hollow spaces. In the way of teeth, the skull contained four molars in each jaw, three pairs of upper incisors, and one pair of lower incisors. From this dentition, we can deduce that the diprotodons were herbivorous—probably browsers, rather than grazers, as their incisors enabled them to strip vegetation from branches. The molars, with their flat surfaces, ground the food before it was swallowed. In the skeletal remains of some diprotodons found at Lake Callabonna, the remains of saltbush were identified where the stomach would have been. This plant is far from nutritious, and it is likely that they only ate such things when they were starving, for example, during the dry season. So we know these marsupials were browsers, but what effect did this have on their activity; were they energetic creatures, always dashing about on the Australian plains, or did they lead a more sedate lifestyle? If the modern wombats are anything to go by, the diprotodons may have had a very slow metabolic rate, enabling them to make the very best of low-energy plant food. If this were the case, then they probably only moved with any urgency when they really needed to.

Like all marsupials, diprotodons had a pouch. There are even bones of adult female diprotodons that are accompanied by the tiny skeletons of their joeys, which were in the pouch when their mothers died. Marsupial babies are born at a very early stage of development. Little more than embryos, they struggle through their mother's fur to the pouch and latch onto one of the teats inside. The teat expands in their mouth, and they're locked in place for the next few months, swallowing their mother's milk. When the diprotodon baby outgrew the pouch, it ventured out into the wide world, keeping close to its mother and retreating to its furry refuge at the first sign of danger, in much the same way as a kangaroo joey.

Today, Australia is bereft of its large, native land predators, but thousands of years ago, this land was home to several creatures that could have made short work of a young diprotodon that had wandered too far from its mother. There was the marsupial lion, with its formidable claws and teeth, and it is likely that this predator killed and ate young diprotodons, and even the adults of the smaller species. The thylacine was another animal that may have preyed on these big, lumbering marsupials, and although it is unlikely that an individual marsupial lion or thylacine could have overpowered and killed the largest, fully grown diprotodons, these extinct predators may have hunted in groups to bring down prey much larger than themselves. In Tasmania, the hind limb bones of one of the smaller diprotodons were found with partially healed teeth marks, thought to be work of the marsupial lion. Saltwater crocodiles (*Crocodylus porosus*) can still be seen in Australia today, and these giant reptiles must have been more than a match for an adult diprotodon. Thousands of years ago, crocodiles were not the only murderous reptiles capable of preying on diprotodons: the giant monitor lizard, *Megalania*, also stalked the land (see the entry "Giant Monitor Lizard" later in this chapter).

These great, pouched plant munchers are, unfortunately, no longer with us. They disappeared, along with most of the Australian megafauna, around 30,000 to 40,000 years ago, just before the peak of the last ice age. The reason for the disappearance of Australia's large mammals is a mystery, but the widely held theory today is that they succumbed to a combination of climate change and human activity. Several thousand years before they became extinct, the earth's climate cooled significantly, and Australia's arid interior expanded to cover over 70 percent of the continent. The diprotodons needed a lot of greenery to sustain their considerable bulk, and as this died back and the standing water disappeared, they may have slowly perished. The ancestors of the Australian Aborigines first reached Australia at least 50,000 years ago; there is little direct evidence of humans hunting the diprotodon, but there is evidence from the Cuddie Springs site suggesting that people may have scavenged from the carcasses of these animals or ambushed them at water holes. It has also been suggested that humans altered Australian habitats by starting bushfires as a way of clearing land or driving prey from cover. However, of the 69 species of extinct Australian mammal known today, only 13 are known to have lived within the period of human occupation. Perhaps people just accelerated a process that had started well before they arrived.

+ It has been suggested that the diprotodons were semiaquatic like the hippopotamus, spending most of their time in lakes and rivers, browsing on aquatic vegetation.
+ The fact that Australian Aborigines lived alongside diprotodons in some parts of Australia for thousands of years is the reason why some people believe that the stories of the *bunyip*, a terrible aquatic beast, are based on folk memories of living diprotodons. The *bunyip* is said to be a dangerous animal that will kill any creature that ventures into its aquatic home. However, it is often the case that due to huge stretches of time, the recollections of extinct animals that persist in folk memories are often massively distorted.

Further Reading: Wroe, S., M. Crowther, J. Dortch, and J. Chong. "The Size of the Largest Marsupial and Why It Matters." *Proceedings of the Royal Society of London B: Biological Sciences* 271 (2004): S34–S36; Wroe, S., and J. Field. "A Review of the Evidence for a Human Role in the Extinction of Australian Megafauna and an Alternative Interpretation." *Quaternary Science Reviews* 25 (2006): 2692–2703.

AUSTRALIAN THUNDERBIRD

Scientific name: *Dromornithids*
Scientific classification:
  Phylum: Chordata
  Class: Aves
  Order: Anseriformes
  Family: Dromornithidae
When did it become extinct? The last of the Australian thunderbirds died out around 30,000 years ago.
Where did it live? The bones of these birds are known only from Australia.

Today, Australia is home to two species of giant flightless bird: the emu of the bush and plains and the cassowary of the northern forests. These two species are closely related to the other ratites, the giant flightless birds that evolved on the immense southern landmass of Gondwanaland: the ostrich of Africa, the rhea of South America, the kiwis of New Zealand, and the extinct moa and elephant birds of New Zealand and Madagascar, respectively.

Copyright Rod Wells

Australian Thunderbird—Stirton's thunderbird (*Dromornis stirtoni*) was probably the largest Australian thunderbird and one of the heaviest birds ever to have lived. (Rod Wells)

Up until 30,000 years ago, Australia supported even more types of giant flightless bird, which were very distinct from the ratites. Collectively, these feathered brutes are known as the dromornithids, or thunderbirds, and they appear to have been diverse, common animals of prehistoric Australia. Seven species of Australian thunderbird have been identified from remains found throughout the continent, and they range in size from animals the size of the cassowary to Stirton's thunderbird (*Dromornis stirtoni*), a 3-m-tall, 400-kg whopper that may challenge the elephant bird, *Aepyornis maxiumus*, for the mantle of the largest bird ever.

The Australian thunderbirds share certain characteristics with the ratites, such as an absent keel bone (the anchor for the attachment of large flight muscles); tiny wings, useless for flying; long legs; and powerful feet. Outward similarities in nature can be misleading, and the parallelism between the thunderbirds and the ratites is simply due to the phenomenon of convergent evolution. The origins of the thunderbirds are very different from the origins of the ratites. Essentially, they were ducks that grew to enormous proportions in the isolated refuge of Australia.

The first bone of a thunderbird was encountered in the late 1820s in the Wellington Caves, New South Wales, by a team led by Thomas Mitchell, but almost 50 more years went by until the first species of thunderbird was formally identified by Richard Owen. Since then, many thunderbird bones have been found throughout Australia. The most common finds have been vertebrae, the long bones of the hind limbs, and toe bones. Bird skulls are particularly fragile, and until very recently, no one had much of an idea how the head of a thunderbird looked. Recent discoveries show that these birds had enormous heads and very impressive beaks. The beaks are very deep, but quite narrow, and some of the species appear to have been equipped with powerful jaw muscles. Naturally, the impressive biting apparatus of these extinct birds has led paleontologists to speculate about what they ate. Some paleontologists believe that they were carnivores, or perhaps even scavengers capable of breaking the bones they found at carcasses. Others believe that the thunderbirds were herbivores fond of nibbling vegetation and using their terrific bill to crack open seeds and nuts. The image of a giant, carnivorous duck is an enticing one, especially for the media, but it is highly unlikely that these huge birds were meat eaters, or even scavengers. They lack the equipment of true predatory animals. The bill may be big, but it certainly isn't hooked, a necessary tool for any animal hoping to tear chunks of flesh from a carcass. Also, the feet of the thunderbirds lack the talons we see in all predatory birds, regardless of their size. Last, the eyes of the thunderbird are not positioned in a way to provide binocular vision: they are situated on either side of the head and give good all-around vision but leave blind spots directly in front of and behind the animal. This is the vision of an animal that is hunted, not a hunter. Chemical analysis of numerous egg shell fragments from one type of thunderbird shows that this species was undoubtedly a herbivore with a penchant for eating grass. Other common thunderbird fossils also point to herbivory. Along with the bones of thunderbirds, paleontologists have unearthed numerous polished stones, known as gastroliths. These were swallowed by the bird and ended up in the gizzard, where they helped break up fibrous plant matter.

As it's very probable the thunderbirds were herbivorous, the numerous predators that once stalked Australia must have hunted some of these birds, especially before they reached adulthood. This is one reason why some of the thunderbirds grew so huge, as large size is an excellent defense against predators. Their other defense was powerful legs, which probably endowed some of the species with a powerful kick and a good turn of speed to get them out of harm's way.

As well equipped as they were to deal with the rigors of prehistoric Australian life, these giant birds lacked the adaptability to deal with the combination of humans and the devastation they bring and climate change. Exactly when the thunderbirds became extinct is a cause of dispute among paleontologists, but the last species is widely thought to have clung to existence until around 30,000 years ago. Scientists have used ancient egg shells of one species of thunderbird (*Genyornis newtoni*) to assess the impact of human activity on these birds, and the Australian landscape in general. It seems that before 50,000 years ago (before the widespread human colonization of Australia), this particular thunderbird pecked at nutritious grasses. However, only 5,000 years later, the diet of this species had completely switched to the leaves of bushes and trees. The scientists' theory is that around 45,000 years ago, humans began to have a drastic effect on the fragile Australian landscape by starting bushfires, which may have burned out of control. With their preferred food up in smoke, the thunderbirds were forced to eat other plant matter, and it seems that they may not have been able to adapt to this change. In the centuries that followed the human colonization of Australia, the thunderbirds dwindled away to extinction.

+ In some Australian Aboriginal rock paintings, there are birds that appear to represent the thunderbirds. The depictions are certainly too large for emus and cassowaries and are probably the artist's attempt at painting one of the larger thunderbird species. Engraved trackways depicting the footprints of large flightless birds have also been found in Australian rocks, and these, too, are thought to represent the thunderbirds.
+ Footprints, thought to be made by the thunderbirds themselves, have also been found in the Pleistocene Dune Sands of southern Victoria, Australia.
+ Apart from the effects of deliberate bushfires, it is very likely that thunderbirds were hunted for food by the first Australians.

Further Reading: Murray, P., and P. V. Rich. *Magnificent Mihirungs: The Colossal Flightless Birds of the Australian Dreamtime.* Bloomington: Indiana University Press, 2003.

GIANT MONITOR LIZARD

Giant Monitor Lizard—The giant monitor lizard was an enormous predatory reptile that prowled ancient Australia up until around 40,000 years ago. (Renata Cunha)

Scientific name: *Megalania prisca*
Scientific classification:
  Phylum: Chordata
  Class: Reptilia
  Order: Squamata
  Family: Varanidae
When did it become extinct? This lizard is thought to have become extinct around 40,000 years ago.
Where did it live? The remains of this animal have only been found in Australia.

Thousands of years ago, Australia was home to more than just giant marsupials. Between 1.6 million and 40,000 years ago, a giant lizard also stalked this fascinating place. Remains of the giant monitor lizard are rare, but enough remnants have been found to allow the entire skeleton of this animal to be reconstructed, and it seems that this was a true giant. The largest living lizard is the Komodo dragon (*Varanus komodoensis*), and in the wild, they can grow to around 3.1 m in length and 166 kg in weight—imagine a bulkier version of the Komodo dragon, which could have been anywhere up to 7 m long and more than 1,000 kg in weight. This is the giant monitor lizard, and its great size alone must have been more than enough to strike fear into the hearts of the first human inhabitants of Australia.

All living monitor lizards are carnivorous animals, and there is no reason to think that the giant monitor was any different. Sections of the animal's jaw have been found, and these prove that this reptile was a meat eater as they are studded with numerous curved teeth. If the size estimates of the giant monitor lizard are true, a fully grown specimen was Australia's largest land-dwelling predator by quite some margin, and there must have been few, if any, animals that it was not capable of tackling.

Perhaps the best way of reconstructing the behavior of the giant monitor is by using the Komodo dragon as a model. This famous monitor lizard has been closely studied for years, and we know a great deal about its general biology. Like the Komodo dragon, the giant monitor probably relied on ambush to catch its prey. It may have skulked in the undergrowth near a watering hole and waited for a hapless victim to come within distance. All lizards, particularly the large monitor lizards, are incapable of maintaining a burst of speed for any significant distance. Their bodies are badly designed for long-distance running as their legs are splayed out to the side and their spines flex from side to side, which makes breathing impossible during energetic movement. In contrast, the legs of a four-legged mammal are directly beneath it, and its spine flexes up and down, which actually helps with breathing (think of a greyhound or cheetah running at full speed). So the giant monitor was limited to lightning strikes from cover, which is still a very effective technique. When the victim was a large marsupial, the giant monitor probably lunged for the neck or the soft underside, which is what the Komodo dragon does when attacking a goat or a water buffalo.

During a predatory attack, the Komodo dragon delivers a bite with its mouthful of teeth and makes no effort to cling on to its terrified prey. This is because the lizard has potent weapons: venom and saliva swarming with bacteria. A bite from a Komodo dragon usually causes a fatal infection, and the victim dies after a few days. With its powerful sense of smell, the lizard follows the scent of death to the final resting place of its prey. Recent research suggests that many types of monitor lizard are slightly venomous, and the giant monitor lizard may have been no different. In actual fact, Australia is home to a bewildering diversity of very venomous animals, and perhaps the giant monitor's saliva was poisonous as well as teeming with dangerous bacteria. We have no way of knowing for sure if this is how this lizard dispatched its prey, but the image of a 7-m-long lizard tasting the air with its big forked tongue, searching for the scent of the doomed animal it has just bitten, is a tantalizing one.

As the giant monitor lizard was so large, it could probably survive on very little food, perhaps only needing to feed once every month, or even less. However, when hunger started to bite and an attack ended in a kill, the giant monitor could have eaten a huge amount of food in one go. In a single meal, the Komodo dragon can gorge 80 percent of its own body weight in food, which is made possible by its very stretchy stomach. The Komodo dragon is also very indiscriminate when it's tearing at the dead body of its victim and everything is eaten. All the indigestible parts, that is, hair, teeth, horns, and so on, are regurgitated after digestion as a pellet smothered in foul-smelling mucus. It is possible that the giant monitor also regurgitated a stinking pellet, but on a much larger scale.

The giant monitor lizard was probably the top Australian predator, but the position of top predator in a food chain is a very precarious one as any big changes in the environment will be felt most powerfully in these lofty reaches. Something did happen around 40,000 years ago that toppled the giant monitor and many other unique Australian species. Due to global cooling, the climate of Australia is thought to have become much drier, and as the rainfall patterns changed, the vegetation began to adapt to the new climatic regimes, and much of Australia became the arid landscape we know today. As the vegetation changed, the populations of the large marsupial herbivores started to dwindle and vanish, until the giant monitor lizard had nothing left to eat. Humans may have encountered this giant lizard, and it must have been a source of wonder and fear. There is long-standing theory that humans changed the face of Australia by starting wildfires. If this occurred, the large-scale burning not only deprived the lizard's prey of food, but may have also killed the reptiles themselves and destroyed their nests.

+ The giant monitor lizard is a favorite of cryptozoologists who believe that this reptile still haunts the Australian outback. There have been numerous sightings that people attribute to this lizard, some of which have been reported by very credible witnesses. It is worth remembering that Australia is a huge, sparsely populated place. A startling example of just what secrets this place still holds is the Wollemi pine (*Wollemia nobilis*), which was discovered in 1994. This living fossil had clung to existence in some remote canyons in the Blue Mountains. If a static species, such as a tree, can remain undetected during two centuries of scientific endeavor, then what are the chances of a highly mobile, albeit giant lizard, still being at large in the Australian wilderness?

+ The larger monitor lizards spend almost all of their time on the ground, but they are proficient climbers and excellent swimmers. However, when young they prefer to spend their time in the trees as they are a tasty morsel for lots of predators, including adults of their own species. Young giant monitor lizards may have spent their early youth in the trees, well out of the way of their enormous relatives.

Further Reading: Molnar, R. *Dragons in the Dust: The Paleobiology of the Giant Monitor Lizard Megalania*. Bloomington: Indiana University Press, 2004; Wroe, S. "A Review of Terrestrial Mammalian and Reptilian Carnivore Ecology in Australian Fossil Faunas, and Factors Influencing Their Diversity: The Myth of Reptilian Domination and Its Broader Ramifications." *Australian Journal of Zoology* 50 (2002): 1–24.

QUINKANA

Scientific name: *Quinkana fortirostrum*
Scientific classification:
    Phylum: Chordata
    Class: Reptilia
    Order: Crocodilia
    Family: Crocodylidae
When did it become extinct? The most recent *Quinkana* remains are around 40,000 years old.
Where did it live? *Quinkana* was an Australian reptile.

Quinkana—*Quinkana* was one of a number of unusual, land-dwelling crocodiles that roamed Australia and the islands of the South Pacific for many millions of years. (Renata Cunha)

In 1970, a caver exploring Tea Tree Cave in north Queensland, Australia, discovered part of the skull of a reptile lying upside down on the cave floor about 60 m from the entrance. Realizing that the skull was something special, she reported her find, and paleontologists returned to the cave to take a look at the skull. The caver had stumbled across the remains of a long-dead, land-dwelling crocodile that was later described and given the name *Quinkana*.

These reptiles, known as mekosuchine crocodiles, are known only from Australia and the South Pacific, and all of them are extinct. The crocodilians with which we are familiar are all amphibious animals that spend nearly all their time in or near water. They are excellent swimmers and remain submerged for long periods of time; however, on land, they can be quite lumbering. The legs of a crocodile splay out to the sides of the large body, and as a result, they are not very effective at supporting the reptile's weight. Also, legs that sprout from the side of the body are not very good when it comes to long-distance walking or running. To make any progress on land, a crocodile moves in a snakelike fashion, with its spine flexing in a horizontal plane, allowing its limbs to gain ground. This movement squeezes the lungs, and if the reptile moves at anything more than a walking pace, it quickly becomes breathless.

The body plan of the *Quinkana* was very different from that of living crocodiles. No limb bones of this animal have ever been found, but similar, yet more ancient crocodiles had relatively long legs that were able to support more of the animal's weight. This arrangement was much better suited to a life on land compared with the crocodiles we know today. It's doubtful that these reptiles were capable of high-speed, long-distance pursuits, but over short distances, they must have been quite deadly.

The fossil record of *Quinkana* is not fantastic, but from the remains we do have, it is possible to estimate the size of this beast—estimates run from 2 m all the way up to 5 m—but the living animal was probably around 3 m long. A 3-m-long, terrestrial crocodile must have been quite an animal and surely an effective predator. The crocodiles are all meat eaters, and *Quinkana* was no different. However, unlike today's crocodilians, which have conical teeth, *Quinkana* jaws were lined with lots of curved, bladelike teeth that were effective tools for slashing at prey.

Exactly what this reptile hunted and how it hunted them is a mystery, but the Komodo dragon gives us valuable insight on the hunting and feeding behavior of a giant reptile. *Quinkana's* Australia was a very different land to the place we know today. The Australian megafauna—a myriad of extinct beasts, some of them huge—once roamed this southern landmass, and many of these animals were fair game for *Quinkana*. Marsupials like diprotodons—giant, wombatlike animals—fell prey to this crocodile. Although *Quinkana* was better adapted for a life on land than the crocodilians we know today, it probably still spent a good deal of its time near sources of water as these attracted large numbers of herbivorous marsupials and other animals on which this reptile could have preyed, including giant birds. The *Quinkana* probably used ambush tactics to surprise its prey. Using undergrowth as cover, the crocodile may have stalked to within striking distance of its victim using its excellent sense of smell and then, when its quarry was within range, it burst from cover with an explosive turn of speed. Lunging at the prey with its mouth open, the jaws snapped shut on the victim.

Many of the modern crocodile species can take very large prey; they do this by dragging the unfortunate animal into the water and drowning it. *Quinkana* was more of a landlubber, and killing large animals without the advantage of water was probably very difficult. If it latched its powerful jaws onto the hind leg of something like a diprotodon, it may have found itself in serious trouble as an enraged, 3-tonne marsupial would have been able to inflict serious injury on a 250-kg reptile. With this limitation in mind, perhaps *Quinkana* had to be content with preying on smaller animals that were killed with a simple snap of the jaws, or with hamstringing larger prey and tracking them to their deaths, a similar technique to that employed by the Komodo dragon.

The most recent *Quinkana* remains are around 40,000 years old, and as is the case for most extinct animals from this period, we have no accurate idea of exactly when this species died out. It may have been around up until very recent times, but until we find the bones, we'll never know for sure. Australian Aborigines undoubtedly came face-to-face with *Quinkana*, and unfortunate individuals may have even fallen prey to it. To what extent humans hunted this reptile, if at all, is unknown, but such a large, land-dwelling animal may have been hunted by humans at some point in the past.

We do know that the most recent bones of this animal come from a time in Australia's history that is marked by the disappearance of many of its amazing animals. Around this time, global cooling was gripping the planet, and although Australia was never buried beneath ice, weather systems the world over were affected. Rains failed, and Australia dried out. Humans may have also modified the habitats of Australia by starting bushfires to clear undergrowth. In combination, climate change and human activity caused the Australian vegetation to die back, and the herbivores began to disappear as their food dwindled. With prey becoming scarcer and scarcer, predators like *Quinkana* were also hit hard, and they, too, eventually became extinct.

- There were once several species of mekosuchine crocodile living in Australia and the South Pacific. The remains of these animals have been found on numerous islands in the South Pacific, but they probably didn't get to these islands by swimming as it is thought that they had no tolerance to saltwater. Perhaps, like smaller reptiles, they were carried between the islands on rafts of vegetation that were broken away by storms and floods. It is thought that Vanuatu and New Caledonia were probably the last refuges of these reptiles, and it is very likely that they survived on these islands until the arrival of humans in quite recent times.

+ Other mekosuchine crocodiles, close relatives of *Quinkana*, have also been discovered in Australia. Some of these remains are around 24 million years old, which shows that *Quinkana* and its relatives were a successful group of animals.
+ Over the last 50 million years or so, at least five other groups of crocodiles have stalked the land, and for a while, some of them competed with mammals in North America and Asia for the supremacy of terrestrial habitats following the demise of the dinosaurs.
+ The name *Quinkana* comes from the Aboriginal word *quinkan*. To some of the indigenous people of Australia's Cape York Peninsula, *quinkans* are humanoid spirits that live in caves and other dark places.

Further Reading: Molnar, R. E. "Crocodile with Laterally Compressed Snout: First Find in Australia." *Science* 197 (1977): 62–64.

GIANT SHORT-FACED KANGAROO

Scientific name: *Procoptodon goliath*

Giant Short-Faced Kangaroo—The grapple-hook paws and the single hind claws of this enormous kangaroo can clearly be seen in this illustration. (Phil Miller)

Scientific classification:
    Phylum: Chordata
    Class: Mammalia
    Order: Diprotodontia
    Family: Macropodidae

When did it become extinct? This kangaroo became extinct around 40,000 years ago.
Where did it live? The giant short-faced kangaroo was found only in Australia.

An enduring image of the Australian wildlife has to be a kangaroo with a cute joey emerging from its pouch. Kangaroos are the quintessential Australian mammals. Among the most familiar of all the marsupials, they have adapted to almost all the habitats the Australian continent has to offer, including open plains, forests, rocky outcrops, slopes, and cliffs. There are even tree-dwelling Kangaroos. These marsupials have a distinctive body shape: a stout body, massively enlarged hind limbs, and a long, muscular tail.

Lots of animals hop, but the kangaroos are the largest animals to use hopping as their preferred mode of locomotion. The kangaroo's hop is actually a very efficient means of getting around as it requires very little muscular effort at moderate speeds. The tendons that stretch down the back of the hind legs to the hugely elongated feet act like springs, and when the animal has gained momentum, these springs help supply much of the power for the hop. Like the limbs of the fleet-footed placental mammals, for example, horses, which end in a single hoof, the digits on the hind limbs of many kangaroos are reduced, and only one of them, the fourth toe, may be in touch with the ground, thus minimizing friction. The large tail acts like a counterbalance at high speed and as a prop to support the body weight of the animal when it's moving about slowly, foraging.

As well adapted as they are, the kangaroos have not escaped the devastation that has seen the extinction of numerous Australian marsupials. Of the 53 species of kangaroo and their close relatives that existed when Europeans first reached Australia, six have become extinct. If we go even further back, into the late Pleistocene, there were many more species, all of which have since died out. The largest living kangaroo by quite some margin is the male red kangaroo, which can stand around 1.8 m tall and weigh in the region of 90 kg. We have seen how the mammals from thousands of years ago were far larger than their extant relatives, and the kangaroos are no different.

The giant short-faced kangaroo was a big marsupial. In life, it probably weighed in the region of 200 kg and reached a height of 2 m. Unlike the largest living kangaroos, this extinct giant had a large, koalalike head with eyes that were more forward facing than those of living kangaroos and hands with long, central fingers, resembling grappling hooks, instead of normal paws. The feet of this hopping brute were reduced to a single, large fourth toe tipped with a single hooflike nail. With such a small surface area in contact with the ground, the animal could hop around the open forests and plains of Australia with considerable efficiency. All the large living kangaroos are dedicated herbivores, and we can safely assume a plant-based diet for the short-faced kangaroo. Its koalalike head suggests a leaf-eating habit. Perhaps it used the grappling hooks on its forepaws to bring high tree branches to within reach of its mouth to nibble the leaves. Marsupials, like all mammals, cannot digest plant matter without the help of symbiotic micro-organisms. Animals like cattle have a chambered stomach that allows plant food to be broken down by the micro-organisms. Kangaroos have a similar system, and most of their micro-organisms are to be found in the first chamber of their complex stomach.

Although the giant kangaroo was undoubtedly a herbivore, it is difficult to explain why it had forward-facing eyes. Living kangaroos' eyes are on the sides of their heads, giving them a 300 degree field of view, excellent for spotting predators. Perhaps the giant kangaroo was simply too big for the Australian predators to tackle and therefore had no need for a wide field of view. There were once numerous large predators in Australia, and only adult giant kangaroos may have had some protection from these animals because of their size. Forward-facing eyes gave the giant kangaroo a good degree of binocular vision and a better perception of distance than kangaroos with a wide field of view. This could be very important for an animal that was moving at high speed through areas of open forest and tall shrubs, where there were numerous obstacles to negotiate. It may have also helped when reaching up into trees to select the most nutritious leaves. With that said, large herbivores are suited to surviving on low-quality food, and the forward-facing eyes may have given the living animal an advantage we will never fully understand.

The giant kangaroo bounded around the wilds of Australia for a long time. The oldest fossils of this animal are around 1.6 million years old, whereas the most recent are 40,000 years old. It seems to have died out at around the same time as the majority of the Australian megafauna. Unfortunately, the definitive explanation for the extinction of these animals is elusive. There are some scientists who believe that the first human inhabitants of Australia are solely to blame, while there is another group of experts who think that climate change was responsible. As we have seen, prehistoric extinctions can very rarely be attributed to a single cause, unless the landmass in question is a small island. In the majority of cases, the evidence indicates a number of causes in combination ultimately leading to the extinction of a large number of species. The probable causes for the disappearance of the giant kangaroo were the spread of humans through Australia and climate change. Humans modified the landscape through their use of fire and probably hunted the giant kangaroo. Climate change made this continent more inhospitable to the large animals, which are often more sensitive to environmental change.

+ The giant kangaroo was not closely related to the group that contains the large, living kangaroos. Its closet living relative is the banded hare-wallaby (*Lagostrophus fasciatus*), a small animal, barely 2 kg in weight, that is extinct on the mainland.
+ The group to which the giant kangaroo and the banded hare-wallaby belong is known as the sthenurinae (Greek for "strong tails"). This group of marsupials diversified about 2 million years ago, and it was once represented by numerous species, all of which are now extinct, apart from the banded hare-wallaby. The giant short-faced kangaroo was the largest, but many of the other species were also very large, far bigger than the living red kangaroo.
+ The bones of the short-faced kangaroo have been found in many sites across Australia, including the Naracoorte World Heritage fossil deposits in South Australia.

Further Reading: Helgen, K. M., R. T. Wells, B. P. Kear, W. R. Gerdtz, and T. F. Flannery. "Ecological and Evolutionary Significance of Sizes of Giant Extinct Kangaroos." *Australian Journal of Zoology* 54 (2006): 293–303.

GIANT ECHIDNA

Giant Echidna—About the same size as a sheep, the giant echidna would dwarf its living relatives. (Phil Miller)

Scientific name: *Zaglossus hacketti*
Scientific classification:
    Phylum: Chordata
    Class: Mammalia
    Order: Monotremata
    Family: Tachyglossidae
When did it become extinct? The giant echidna died out about 40,000 years ago.
Where did it live? The remains of the giant echidna have only been found in Australia,
    but its range may have included New Guinea.

The monotremes are a very odd group of mammals that have perplexed zoologists for decades. In some ways, they are unquestionably mammals as they have fur, nourish their young with milk, and are able to keep their body temperature constant by metabolizing food. However, they also have some reptilian features, that is, they lay eggs and their feces, urine, and eggs emerge from a common opening: the cloaca. The first species of monotreme to come to the attention of European scientists was the platypus (*Ornithorhynchus anatinus*) when the dried skins of this animal were sent to England from Australia. These skins caused uproar among the zoological fraternity. There were cries of fake! and sham! as many experts of the time claimed it to be nothing more than the abominable creation of a mischievous taxidermist. Gradually, scientists accepted that the platypus was a living, breathing animal and not the work of an imaginative taxidermist. Not long after the platypus came to the attention of Europeans, the echidna was described and named by scientists.

Today, four species of echidna have been described, and all of them, more or less, bear a superficial resemblance to hedgehogs. They have long spines on their back, and their small head ends in a thin snout. Three of the living echidna species have a relatively long snout and are known as long-beaked. Long-beaked echidnas are known only from the highlands of New Guinea. The most common living species is the short-beaked echidna, and it is found all over Australia and in some parts of New Guinea. The short-beaked echidna is a specialist predator of ants and termites. It probes the ground and insect nests with its long snout and uses its long tongue to bring the prey to its mouth, in much the same way as an anteater; indeed, another, albeit incorrect name for the echidnas is "spiny anteaters." The long-beaked echidnas are similarly equipped with a long tongue, but theirs is equipped with spines for extracting earthworms from the soil.

The echidnas are a specialty of Australasia and are only known from Australia and New Guinea. As with any group of living mammal, these odd animals were once represented by giant species, and up until 40,000 years ago, there lived an echidna as large as a sheep. Today, the largest echidna species is the western long-beaked echidna (*Zaglossus bruijni*), which can weigh as much as 16 kg; however, the giant echidna was about 1 m long and weighed at least 50 kg. All the living echidnas are specialist predators of invertebrates, so we can be confident that the giant echidna was no different, although it is impossible to tell if this extinct monotreme preferred to eat ants or worms.

The giant echidna's large size may have afforded it protection from some predators, but the thylacine, marsupial lion, and giant monitor lizard were all large enough to tackle an echidna, albeit a giant one. It is therefore likely that the giant echidna was protected with spines in the same way as the living species. The echidna's spines are actually individual hairs, and they are rooted in a layer of thick muscle, which covers the whole body—the panniculus carnosus. When a short-beaked echidna feels threatened, it pulls its legs and head under its body and erects its spines. The potential predator is met with a bristling ball of spines, and after a few minutes of getting spiked in the face and paws, it often gives up and leaves the echidna alone. On soft ground, the short-beaked echidna can enhance its defense still further by burrowing into the ground until only a crown of spines can be seen. It takes a very determined predator to beat the echidna's defenses.

Like all the other monotremes, the giant echidna must have laid eggs. Unlike a marsupial, an echidna's pouch is not well developed. Outside of the breeding season, the pouch is nothing more than a groove on the female's belly, but hormonal changes around the breeding season cause the groove to become more well developed, until there is a shallow pouch in the female's abdomen. After mating, the female echidna everts her cloaca and deposits a single rubbery egg with a diameter of 13 to 17 mm into the simple pouch. After about 10 days, the young echidna (puggle) hatches, but there are no teats for it to latch on to; instead, it grips a special patch of milk-producing skin at the front end of the pouch. It laps at the pinkish milk and stays in the pouch for 2 to 3 months, until the mother has to turn it out because of its growing spines.

Like the short-beaked echidna, the giant echidna probably ranged over much of Australia, but it appears to have been another casualty of the changes that affected Australia 40,000 years ago, and it became extinct with almost all of Australia's varied megafauna. Some experts suggest that climate change was the major cause of these extinctions, but others think that the first human inhabitants of Australia annihilated the native fauna through hunting and habitat destruction. It's very likely a combination of these factors that led to the disappearance of these creatures.

- The giant echidna was described from an incomplete skeleton found in Mammoth Cave, Western Australia. A second species of huge echidna has been described from a 65-cm-long skull, so it is likely that the prehistoric Australia was home to at least two species of very large monotreme.
- Echidnas have very large salivary glands, and these secrete a thick, sticky saliva that lubricates the tongue as it's protruded in and out of the mouth. The saliva also traps the echidna's prey.
- The monotremes are a very ancient group of mammals. The oldest known monotreme fossil is an opalized lower jaw fragment from the Lightning Ridge opal fields of New South Wales, which is around 100 million years old. A fossilized platypus tooth has even been found in Argentina, demonstrating that these animals have not always been restricted to Australia and New Guinea. It is highly likely that monotremes were once found all over the ancient landmass of Gondwanaland.

Further Reading: Griffiths, M., R. T. Wells, and D. J. Barrie. "Observations on the Skulls of Fossil and Extant Echidnas (Monotremata: Tachyglossidae)." *Australian Mammalogy* 14 (1991): 87–101; Pledge, N. S. "Giant Echidnas in South Australia." *South Australian Naturalist* 55 (1980): 27–30; Murray, P. F. "Late Cenozoic Monotreme Anteaters." *Australian Zoologist* 20 (1978): 29–55.

WONAMBI

Scientific name: *Wonambi naracoortensis*
Scientific classification:
   Phylum: Chordata
   Class: Sauropsida
   Order: Squamata
   Family: Madtsoiidae
When did it become extinct? This snake became extinct around 40,000 years ago.
Where did it live? This snake was only found in Australia.

The snakes are a very odd group of reptiles. Sinuous and legless, they have evolved some amazing ways of catching their food and protecting themselves. Although these limbless reptiles are endlessly fascinating, their origins are nothing short of a mystery. It is thought that their closest relatives are the monitor lizards, and although snake fossils are quite common, they can be hard to study, so we can only guess at how and why these remarkable reptiles evolved from lizard ancestors with functional limbs to the serpents we know today.

The evolutionary history of the snakes may be sketchy, but some answers have been discovered in the home of animal anomalies: Australia. In various cave sites in Australia, paleontologists have found the bones of a long-dead animal that belonged to a very ancient group of snakes, all of which are now extinct. This group, the *Madtsoiidae*, survived for around 90 million years, from the middle of the Cretaceous to the Pleistocene. They were once found in Australia, South America, Africa, Madagascar, and Europe, but they slowly died out, until

Wonambi—The remains of a *Wonambi* lie on the floor of the Naracoorte Caves in South Australia. (Rod Wells)

Australia was their last refuge, and *Wonambi* was one of the last of their number. The bones of this animal from the Victoria fossil caves show that it was a large snake, perhaps as much as 6 m long, which is comparable to some of the largest pythons and boa constrictors alive today.

Like the living giant snakes, *Wonambi* was probably nonvenomous, instead relying on ambush tactics and its muscular body to catch and suffocate prey using constriction. Constriction is actually a very effective means of subduing prey and is used by a large number of snakes, not only by the large boas, pythons, and anacondas. *Wonambi* probably loitered around watering holes and other places that attracted its prey. If a suitable victim came within striking distance of the *Wonambi*'s hiding place, the snake launched a lightning-fast lunge, snagging the prey with its sharp, curved teeth. In the blink of an eye, *Wonambi* threw coil after coil of its long body around the struggling victim. The embrace of *Wonambi* must have been an inescapable one as the reptile tightened its grip, slowly suffocating the victim with crushing force. When the prey was dead, *Wonambi* relaxed its grip and set about swallowing the still warm body. The snakes we know today have a fantastically flexible skull and lower jawbone that makes it possible for them to swallow large animals. Large pythons and the anaconda can inch their head over their prey until the whole body is engulfed. *Wonambi* was a primitive snake, and it lacked the highly flexible skull of the modern snakes; therefore it was probably limited to smaller prey such as the many species of smaller wallaby that still inhabit Australia. The larger marsupials, many of which are now extinct, were probably too big for *Wonambi* to handle, but any animal visiting a water hole in ancient Australia was probably always wary of being caught in a *Wonambi* ambush.

As the living giant snakes can catch and eat huge prey animals, they can go for many months between meals. They rest and digest their prey for several days or weeks, and their very efficient metabolism enables them to make the very most of all the food they eat. As *Wonambi* didn't have the head or jaws for large prey, it may have needed to eat more frequently than the living constrictors.

Wonambi was the last of a long line of primitive snakes and one of many such giants that once slithered their way around Australia. They seem to have died out with the rest of the Australian megafauna around 40,000 years ago, but as new evidence comes to light, this date may change significantly. Humans may have known these snakes, and it is possible that human activities, such as bushfires, led to their demise. Australia, like the rest of the world, has been through some massive climatic changes in the past 2 million years or so, and perhaps the demise of these snakes coincided with the disappearance of the lush vegetation that once shrouded the Australian continent, leaving the arid landscape we know today. Water holes and other habitats favored by *Wonambi* disappeared, and its prey grew increasingly difficult to find. Confronted by this changing world and the pressure of human hunting, the *Wonambi* and the other primitive snakes eventually disappeared.

+ For a long time, it was assumed that the snakes descended from a burrowing ancestor that took to a life underground and lost its limbs. This may be partially true as the eyes of snakes are unique among the vertebrates, with many features that are not seen in any fish, amphibian, reptile, bird, or mammal. Some scientists have argued that this is because the ancestors of snakes were subterranean animals that completely lost their eyes as well as their limbs. As they moved back onto the surface to fill vacant niches, their eyes reevolved into the unique structure we see today.

+ Snake fossils can be numerous, especially the very durable vertebrae. There are even several skeletons of extinct snakes that are more or less complete. Some of the primitive extinct snakes even had hind legs.

+ The vestiges of these hind legs can be seen in the most primitive of the living snakes: the large constrictors (boas, pythons, and the anaconda). In some of these species, the male has a pair of tiny spurs on the back end of his body, which are used during mating. These spurs are the last vestige of the snake's hind limbs. If you were to cut one of these snakes open, you would see the pelvic bones and the vestigial leg bones.

+ The *Wonambi* takes its name from one of the "rainbow snakes," the mythical serpents in the creation stories of the Aboriginal people. Perhaps these myths are based on reality?

Further Reading: Scanlon, J.D., and M.S.Y. Lee. "The Pleistocene Serpent *Wonambi* and the Early Evolution of Snakes." *Nature* 403 (2000): 416–20; Scanlon, J.D., and M.S.Y. Lee. "The Serpent Dreamtime." *Nature Australia*, summer 2001; Brown, S.P., and R.T. Wells. "A Middle Pleistocene Vertebrate Fossil Assemblage from Cathedral Cave, Naracoorte, South Australia." *Transactions of the Royal Society of South Australia* 124 (2000): 91–104.

♀ Extinction Insight: A Hole in the Desert—The Nullarbor Plain Caves

Immediately north of the Great Australian Bight, the large open bay on the south coast of Australia, lies the Nullarbor Plain, the largest single outcrop of limestone in the world, with an area of around 200,000 km². The limestone of this plain was laid down millions of years ago in a shallow sea, but geological activity forced the huge slab into its present position. This flat and treeless semiarid plain is far from inviting, but beneath its surface are treasures.

Limestone is dissolved slowly by rainwater, and over millions of years, any large area of this rock soon becomes riddled with caves and tunnels. This is exactly what has happened to the Nullarbor Plain, and its flat surface belies a network of caverns and tunnels, only a tiny proportion of which have been explored. In May 2002, a group of cavers found a sinkhole on the surface of the Nullarbor Plain—a sinkhole appears when the roof of a limestone cavity is dissolved, leaving a short pipe into the cavern beneath. They decided to explore the sinkhole and lowered themselves through the 11-m pipe and into the cavern below. It was a further 20 m to the floor of the cavern, and when they shined their head torches on the rocks around their feet, they were met with a site that no human had ever before seen. Around them, littering the floor of the cavern, were numerous skeletons. Some of the bones were semiconcealed by sediment, while others were lodged between rocks and boulders. Obviously, these were not the remains of creatures that had died recently, and realizing the importance of their find, the cavers alerted the authorities. Following the discovery, a team of paleontologists and geologists visited the site and lowered themselves

The Nullarbor Plain Caves—A caver is shown descending through a narrow sinkhole into one of the Nullarbor Plain caves, which yielded an unparalleled haul of ancient animal remains in an incredible state of preservation. (Clay Bryce)

into the caves. It soon became clear to them that this was a very important find, probably one of the most significant paleontological discoveries on Australian soil. What lay before them was a more or less complete record of the animals that once stalked the Nullarbor Plain above their heads. The remains of the animals were perfectly preserved, but they were fragile, and before any of the bones were removed, they were painted with a special strengthening compound.

In total, 69 species of vertebrate were identified from the caves, many of which survive in Australia to this day. Twenty-one of the identified animals did not survive the Pleistocene and are known only from bones. Of the 23 species of kangaroo identified from the remains in the cave, no fewer than 8 were new species, which goes to show just how diverse Australia's large animal fauna once was. One of the most interesting of these extinct kangaroos was a small species with bony protrusions above its eyes, like small horns. Exactly what these were for is a mystery, but paleontologists have speculated that they protected the animal's eyes from the spines of its food plants. Interestingly, two of the extinct kangaroos from the cave were tree-dwelling species, similar to the surviving rainforest kangaroos of New Guinea. The site also yielded no fewer than 11 complete skeletons of the marsupial lion, an animal that was only previously known from a handful of skeletons. The largest animal in the assemblage was the extinct giant wombat, *Phascolonus gigas*, which, at 200 kg, goes to show that the sinkhole was quite some pitfall trap.

All of the amazing animals the scientists discovered fell through the opening of the sinkhole and ended up on the chamber floor some 30 m beneath the surface. Not many of the remains were found directly beneath the opening in the chamber ceiling, so it seems the fall was not fatal. Badly injured on the floor of the cave, the hapless animals crawled away into the darkness and died a slow death from their injuries and a lack of food and water. The scientists were finding the animals in the same positions in which they had died thousands of years previously, but when exactly had these animals fallen into the cave? Analysis of the sediments in the cave show that the cavern was first opened to the surface about 790,000 years ago. Over millennia, natural processes had sealed the sinkhole on numerous occasions, only for heavy rain and geological activity to open it again. It seems that the opening closed for the last time about 195,000 years ago, so what's preserved at the bottom of this cave is a 600,000-year record of the animals that once lived in this part of Australia.

These caves show not only how diverse the Australian megafauna was, but also what the landscape and climate were like. Today, the Nullarbor Plain is a relatively lifeless landscape, and the flora of the area is dominated by saltbush (*Atriplex* sp.) and bluebush (*Maireana aphylla*) scrub. Thousands of years ago, this was not the case, as trees and other plants, many of which have since disappeared, were common. Instead of the arid steppe we find today, the Nullarbor Plain was probably a mosaic of woodland and scrub, with plants that bore palatable leaves and fleshy fruits. The fact that arboreal kangaroos have been recovered from the caves is proof that these plains supported large trees thousands of years ago. Interestingly, the climate of the ancient Nullarbor Plain was no different to what we see today, with average annual rainfall of around 180 mm. The drying of Australia's climate is often cited as the cause of the extinction of the megafauna this huge island once supported. The Nullarbor caves suggest otherwise. The animals and plants of Western Australia were well suited to arid conditions, and the disappearance of the bizarre beasts from this arid plain may be due to wildfires (natural or caused by humans) that wiped out many of the plant species, leaving the impoverished landscape we see today. With their food dwindling, the herbivores of the Nullarbor died out, closely followed by the predators and scavengers.

Further Reading: Prideaux, G. J., J. A. Long, L. K. Ayliffe, J. C. Hellstrom, B. Pillans, W. E. Boles, M. N. Hutchinson, R. G. Roberts, M. L. Cupper, L. J. Arnolds, P. D. Devine, and N. M. Warburton. "An Arid-Adapted Middle Pleistocene Vertebrate Fauna from South-Central Australia." *Nature* 445 (2007): 422–25.

7

MORE THAN 50,000 YEARS AGO

GIANT RHINOCEROS

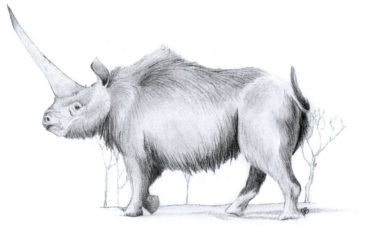

Elasmotherium— This enormous rhinoceros roamed the steppes of Asia. The remnants of its horn have long since disappeared, but in life, this weapon could have been 2 m long. (Renata Cunha)

Scientific name: *Elasmotherium sibiricum*
Scientific classification:
  Phylum: Chordata
  Class: Mammalia
  Order: Perissodactyla
  Family: Rhinocerotidae

When did it become extinct? The most recent specimens of this prehistoric animal are around 1.6 million years old, but there is circumstantial evidence that this great beast survived into much more recent times.

Where did it live? The remains of this animal have been found on the central steppes of Asia and at locations in southern Russia.

Mammoths were not the only giant, shaggy beasts that stalked the cold, windswept lands of central and northern Asia. That other group of massive herbivorous mammals, the rhinoceri, also spawned species that were adapted to the cold conditions that have prevailed on earth for the last 2 million years. The giant rhino was one of these animals. We only know it only from a few skeletons and isolated bones, but even these dry remains are a real sight. The living animal, walking across the treeless plains of central Asia, must have been a very impressive sight. An adult giant rhino was around 6 m long, 2 m at the shoulder, 5 to 6 tonnes, and probably covered in dense fur. By comparison, the biggest white rhino (*Ceratotherium simum*) on record was just over 4 m long and 1.8 m tall, and weighed around 3.5 tonnes, which gives you a good idea of how big the giant rhino was.

The skulls of this animal that have been found indicate that this beast was the proud owner of a single, huge horn, estimated to have been around 2 m long. We can never be sure of the appearance of the horn because one has never been found due to the simple fact that unlike deer antlers, rhinoceros horn is actually made out of very dense keratin fibers, the same protein that makes your hair and nails. In life, these horns are a potent weapon, but in death, they rot away, leaving no remains.

The white rhino, even with its stubby legs, is a quick, nimble runner able to reach speeds of 40 to 50 km per hour, and as the giant rhino had relatively long legs, it may have been capable of quite a turn of speed, with a running gate similar to a horse, characteristics that were very useful on the central Asian steppe. We can be fairly certain that an adult giant rhino was invulnerable to all of the predators of the time, even the saber tooth cats with their huge fangs, but it may have been a different story for young giant rhinos, who were probably easily overpowered and killed by a carnivorous cat or a pack of wolves.

We know from the fossils of the giant rhino that its cheek teeth grew continuously throughout its lifetime, and this gives us insight into what it ate. Like other mammals of enormous bulk, the giant rhinoceros was a herbivore, and it would specifically have favored grasses and the short herbs growing on the steppe. The fibrous vegetation that formed its diet must have been very tough on the teeth, and long hours every day spent chewing wore them down; fortunately, the continual growth of the teeth got around this problem, ensuring that a good grinding surface was always in place to pulverize the plants. Apart from being tough and fibrous, grass is also difficult to digest, and all rhinos, even long-dead ones, employ the help of bacteria to break down the cellulose that forms the bulk of plant tissue into sugars that can be digested. In rhinos, the bacteria process the cellulose in the rear of the gut, which gives them the name "hind-gut fermenters." This type of fermentation is quite inefficient, but it can deal with lots of food in a short period of time, enabling the hind-gut fermenters to reach great size, as the giant rhino did.

Compared to remains of prehistoric mammals like the mammoths, fossils of the giant rhino are very rare, and our knowledge of this amazing, long-dead beast is based on only a few bones that have been unearthed over the years. Of these remains, the most recent ones are around 1.6 million years old, but there is anecdotal evidence that this species survived into much more recent times. Some of the native tribes of central Asia and southern Russian as well as medieval chroniclers tell stories of a great black bull with a single horn on its head. There is no doubt that whatever animal prompted these stories is long extinct, but it is possible that the giant rhino survived for long enough to feature in the folk memory of these people. Some people even suggest that the legend of the unicorn stemmed from the folk memory of the giant rhino, but whatever the truth may be, it is intriguing to think that our ancestors on the lonely plains of central Asia once walked among these gigantic, single-horned rhinoceri.

- The line of mammals that gave rise to the living rhinoceri we know today—the white rhino, the black rhino (*Diceros bicornis*), the Indian rhino (*Rhinoceros unicornis*), the Javan rhino (*Rhinoceros sondaicus*) and the Sumatran rhino (*Dicerorhinus sumatrensis*)—has, over immense stretches of time, been represented by some bizarre and amazing animals, including the largest land mammal ever to have lived: the truly immense *Indricotherium*, which was about 5 m at the shoulder, 8 m long, and 20 tonnes in weight.

- The horn of the giant rhino reflects the exaggeration in reproductive adornments that can be seen in many types of prehistoric mammal—from the giant tusks of the mammoths to the remarkable antlers of the giant deer. The giant rhino's horn was crucial in winning a mate during the breeding season as males could have sized each other up based on the size of their adornment. When fights between males did erupt, the horn must have been a vicious weapon, and it must also have been used with great effect against any predators stupid enough to attack the giant rhino.

- As with the other animals that evolved to survive the cold and warm cycles of the ice ages, the giant rhino was able to cope with the changing conditions that saw global temperatures increase and ice sheets the world over recede, although its populations may have expanded and contracted with the movements of the ice. It is unlikely that human hunting was solely responsible for the extinction of these animals, but it may have been sufficient to knock a species over the edge whose populations were already being squeezed by climate change.

Further Reading: Noskova, N.G. "Elasmotherians—Evolution, Distribution and Ecology." In *The World of Elephants—International Congress, Rome*, 126–28. Rome: 2001; Markova, A.K. "Pleistocene Mammal Faunas of Eastern Europe." *Quaternary International* 160 (2007): 100–11.

MEGATOOTH SHARK

Scientific name: *Carcharocles megalodon*
Scientific classification:
  Phylum: Chordata
  Class: Chondrichthyes
  Order: Lamniformes
  Family: Lamnidae

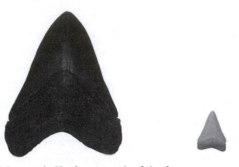

Megatooth Shark—A tooth of the fearsome great white shark, *right*, looks very small indeed next to the tooth of the megatooth shark, *left*. (Ross Piper)

Megatooth Shark—The megatooth shark, *top and center*, was at least 20 times heavier than the living great white shark, *bottom left and bottom right*, making it the largest predatory fish that has ever lived. (Renata Cunha)

When did it become extinct? The megatooth shark is thought to have become extinct around 1.6 million years ago.

Where did it live? This shark appears to have had a global, subtemperate distribution as its fossils have been found in Europe, Africa, North and South America, southern Asia, Japan, Indonesia, Australia, New Caledonia, and New Zealand.

The great white shark (*Carcharodon carcharias*) is one of the most formidable predators in the ocean, yet it would be dwarfed next to the megatooth shark—the largest predatory fish that has ever lived. As its name suggests, megatooth's mouth bristled with an abundance of triangular, serrated teeth that make a great white's dentition look pretty tame.

Sharks and their relatives have a skeleton composed mainly of cartilage, which in life is a very light and flexible frame. In death, however, this frame rots away to nothing as there are no minerals, for example, apatite, that can be replaced by other minerals to form fossils. Due to this quirk of anatomy, all that remains to testify to the existence of this fantastic fish are its immense teeth and disc-shaped parts of the vertebrae known as centra. Many teeth have been found, some of which have been recovered in dredges of sediment from the seabed, while others have been found in quarries in various locations around the globe. The appearance of the shark has been extrapolated from these remains. The teeth can be used to reconstruct the upper and lower jaw, and a body can be built around what must have been a cavernous mouth.

The adult size of this shark is a bone of contention among experts. Some recent calculations estimate the body length of this animal to be 16 m, with a weight of approximately 48 tonnes. By comparison, the largest great white sharks alive today are around 6 m long and 1.9 tonnes. Even these conservative estimates of the megatooth's length and weight suggest a truly terrifying creature that once patrolled the seas of the prehistoric earth. We know that the megatooth was a very large animal, but what did it look like? We can only guess, but for a long time, it was assumed to look like a giant great white. It is now reckoned to have had the same general body shape as the great white, but with a heavier head, more massive jaws, and longer pectoral fins—obviously, these reconstructions must be treated with caution as they based are nothing more than teeth and bits of backbone.

The megatooth shark was undoubtedly a predator, but what did it eat and where did it hunt? The remains that have been found suggest that the shark was an inhabitant of shallow, warm to cool temperate coastal waters—habitats that were commonplace around 10 million years ago. These waters were home to a wealth of marine mammals that had evolved from ancestors that took to the water not long (in geological terms) after the extinction of the dinosaurs. This marine mammal fauna consisted of whales, seals, sea lions, and the extinct relatives of dugongs and manatees. It is probable that the megatooth shark ate all these animals, but it may have been a specialist predator of whales. Fossils of extinct whales have been found bearing deep gashes the right size and shape to have been inflicted by the slashing teeth of the megatooth shark. You can just imagine this 50-tonne shark slamming into the side of an ancient, 10-m-long baleen whale and tearing out a huge chunk of blubber and flesh. Like the great white shark, megatooth probably retired to a safe distance after this initial strike to let the prey bleed to death before closing in to feast. Its food requirements must have been enormous, and if the great white shark is anything to go by, it may have needed about one-fiftieth of its weight in food every two weeks, which, for a fully grown megatooth, was about 1 tonne of meat. An adult megatooth was able to tackle whales, but what did these sharks eat when they were young? They probably fed on large fish and may have had different teeth from the adults, up to the job of keeping a firm grip on slippery fish. The teeth of a young great white are more slender and narrow than those of the adult to provide an advantage in catching fast-moving fish.

Even though the adult megatooth shark must have been the undisputed king of the sea, the great white shark—one of the most impressive predators alive today—actually coexisted with the megatooth. How did these two enormous predatory fish manage to live at the same time without coming into direct competition with one another? They may have managed to coexist by feeding on different prey. As the great white is much smaller than the megatooth, its preferred prey is seals and sea lions, while megatooth was capable of attacking and killing whales. The great white is still around today doing the same thing it has done for millions of years, but all that remains of the megatooth are petrified fragments of its body. What happened to this giant shark?

The megatooth's massive appetite probably made it very vulnerable to the ravages of global cooling, which entered a harsh phase around 2 million years ago. Temperatures at midlatitudes dropped by around 15 degrees Celsius, and as more and more water got locked up in the growing glaciers, megatooth's shallow water habitats became scarcer and colder, and the shark was forced into dwindling pools of habitat, unable to catch sufficient prey to fuel its enormous bulk. Some of the whales on which megatooth probably fed also became extinct at around the same time, supporting the theory that shallow, warm-water habitats disappeared due to global cooling.

- The megatooth shark existed for around 20 million years, and although it is often assumed to be a close relative of the great white, their exact relationship is still uncertain.
- Although adult megatooth sharks were at the very top of the food chain, the young were fair game for many marine predators.
- Sharks have the amazing ability to continually replace their teeth. As a tooth breaks off or is shed, the first in a line of growing replacements moves forward to fill the gap.

For this reason, shark teeth are very common in the fossil record and have been known for centuries—often known by the name of "glossopetrae" (Greek *glosso* translates as "tongue" and *petrae* translates as "stone"). Even Pliny the Elder, the Roman naturalist, wrote about them, believing them to fall from the sky during lunar eclipses. They were later assumed to be serpent's tongues that St. Paul had turned to stone.

+ It has been suggested that the megatooth shark may still survive, but continued survival implies a viable population. In reality, there is no chance that such a huge, surface-dwelling predator could escape detection in the modern age.

Further Reading: Klimley, A. P., and D. G. Ainley, eds. *Great White Sharks: The Biology of Carcharodon carcharias*. San Diego: Academic Press, 1996; Tschernezky, W. "Age of *Carcharodon megalodon?*" *Nature* 184 (1959): 1331–32.

MAGNIFICENT TERATORN

Magnificent Teratorn—The magnificent teratorn was the largest flying bird that has ever lived. At 6 to 8 m, its wingspan was about the same as a small airplane. (Renata Cunha)

Scientific name: *Argentavis magnificens*
Scientific classification:
   Phylum: Chordata
   Class: Aves
   Order: Ciconiiformes
   Family: Teratornithidae

When did it become extinct? The only known remains of this bird are from around 6 million years ago, but we don't have a more accurate idea of exactly when it became extinct.

Where did it live? The remains of this bird have been found in Argentina.

The two species of condor that inhabit the Americas are enormous birds. If you have ever seen one of these birds for real or television footage of one of them tearing at the carcass of a dead animal, you'll appreciate just how big they are. They can be around 1.1 m tall, and their wingspan can be as much as 3.1 m. In the sky, these birds use their huge wings to soar for hours on updrafts of warm air, surveying their immense territories for food. With the living condors in mind, let's travel back in time around 6 million years and visit Argentina. Back then, the Andes were only starting to form due to the tectonic forces that pushed the Pacific plate under the South American plate. As a result, the flat grasslands of Argentina were swept continuously by westerly winds. High above these plains, soaring effortlessly in the sky, was the largest flying bird that has ever lived: the magnificent teratorn. The wingspan of this immense bird was about the same as a small airplane, at 6 to 8 m, and it probably weighed in the region of 80 kg, possibly more. This is really heavy when we consider that the heaviest flying birds today, the great bustard (*Otis tarda*) and the kori bustard (*Ardeotis kori*), are around 20 kg. Standing, the magnificent teratorn was 1.5 to 2 m tall.

Bird skeletons are very fragile, and it is very rare to find an intact one that has stood the test of time. All the vital statistics of this giant have been extrapolated from a few bones found in Argentina. Paleontologists have unearthed some of the wing bones, fragments of the feet, and portions of the skull. Even though we only have fragments, it is possible to piece together a realistic reconstruction of the entire skeleton, and from there, we can build up a picture of how the living animal may have looked and how it may have lived.

The teratorns are related to the New World vultures, for example, the condors. Like the other teratorns, the magnificent teratorn had a hooked bill, so it must have been a meat eater, but how did it go about finding its food? Three plausible ways of life have been proposed for this extinct bird. Some experts have suggested that this bird was an active hunter that swooped down and caught animals as big as hares while on the wing, whereas others believe that it behaved in the same way as the modern-day condor, alighting near a carcass and feasting on the flesh. Another possibility is that this giant spent a lot of time stalking the pampas on foot searching for tasty morsels. After carefully inspecting the skull bones of this bird and its relatives, scientists have proposed that a magnificent teratorn's skull was not really up to the task of tearing the hide and flesh of dead animals. It may have relied on other animals to tear the hide, such as the saber tooth predators, which lived at the same time. These powerful mammals were undoubtedly able to bring down prey much larger than themselves, so there was definitely a source of big, dead animals for a giant scavenger. Perhaps these birds used their immense size to intimidate predators and chase them away from their kill?

Using the information we have on living scavenging birds, it is possible to estimate the size of the territory this giant bird needed to find sufficient food for itself, and it is something on the order of 500 km^2. To survey such a huge territory, the magnificent teratorn must have been on the wing almost continually. Fortunately, a huge wingspan is perfect for effortless gliding on the thermal updrafts that rise up from the pampas. However, there is the one problem of how such a huge bird got airborne if it was on the ground. Massive wings cannot be flapped effectively when you are grounded, and it has been estimated that to get airborne, the teratorn needed to reach a ground speed of 40 km per hour. This is quite fast and beyond the capabilities of the teratorn's feet, which seem to be built for sedate stalking. The solution to this problem could have been the strong, incessant winds that blew across the South American pampas and Patagonia. The magnificent teratorn may just have needed to turn its outstretched wings into the wind, and the speed of the moving air probably lifted it into the sky. It may have also become airborne by running down a slope or dropping from a high perch. The wandering albatross (*Diomedea exulans*), which has the greatest wingspan of any living bird, takes to the air by stretching its wings and running into the wind.

Using what we know about living birds, we can piece together other parts of the magnificent teratorn's life. Such a large bird must have definitely been very long-lived. The living condors can live for at least 50 years, so the extinct giant could have lived to a very old age. Long life is associated with slow breeding, and this huge bird may have only reached sexual maturity after its twelfth year. Once it was capable of producing offspring, it is highly likely that only one chick was reared every two years. Where they constructed their nest and what it looked like is a mystery, but it may have been a simple affair of a few twigs surrounding the 1-kg egg on a substantial cliff ledge that gave the adults sufficient space to take off and land. Great age, slow development, and a low reproductive rate are good reasons for a bird to remain with the same mate for its whole life, and it is an intriguing thought that these giant, long-dead birds, known only from a few bones, formed pair bonds that lasted their entire reproductive life.

It would be a fabulous sight to see a bird of the magnificent teratorn's enormity gliding over the South American pampas and Patagonia, but this animal has long since disappeared from the face of the earth. Its demise cannot be attributed to the changes that occurred at the end of the last ice age, changes that coincide with the disappearance of other American megafauna. We can't attribute its demise to our own species as it disappeared a long time before modern humans arrived on the scene in the Americas. It is likely that as the Andes rose into the air over millennia, the perpetual westerly winds that scoured the pampas were reduced. It is also possible that the strong westerly winds shifted to the south as the postglacial climate changed. Without these strong winds to give them a helping hand into the air, these giant birds may have simply been too large to fly, and over thousands of years, they slowly died out, leaving just fragments of their bodies to provide us with a window to the distant past.

+ Four other teratorn species have been identified, but the species described here is the only one known so far from South America. Bones of two of the other species have been found in great abundance in the asphalt deposits of Rancho La Brea, Los Angeles (see the "Extinction Insight" in chapter 4). The magnificent teratorn was by far the largest of these extinct birds.
+ The teratorns and their living relatives, the New World vultures are more closely related to the storks than they are to other birds of prey. This is another example of convergent evolution, as they have come to resemble the true vultures of the Old World.

◆ As the bones of these giant birds only survive as fragments, it is just a matter of time before more are found and described, giving us a more accurate picture of how these extinct animals looked and behaved.

Further Reading: Paul Palmqvist, P., and S. F. Vizcaíno. "Ecological and Reproductive Constraints of Body Size in the Gigantic *Argentavis magnificens* (Aves, Theratornithidae) from the Miocene of Argentina." *Ameghiniana* 40 (2003): 379–85; Hertel, F. "Ecomorphological Indicators of Feeding Behavior in Recent and Fossil Raptors." *The Auk* 112 (1995): 890–903.

POUCH-KNIFE

Pouch-Knife—A pair of pouch-knife marsupials prepare to go hunting after a long rest. This unusual predator probably used ambush tactics and strength to catch and subdue its prey. (Renata Cunha)

The skull of the pouch-knife clearly shows the huge extensions of the mandible that protected the long canines. The long root of the canines can be seen extending beyond the eye. Very few remains of this animal are known. (Ross Piper)

Scientific name: *Thylacosmilus atrox*
Scientific classification:
 Phylum: Chordata
 Class: Mammalia
 Order: Sparassodonta
 Family: Thylacosmilidae

When did it become extinct? The pouch-knife became extinct around 4 million years ago.
Where did it live? The remains of this animal are only known from Argentina.

Today's land-dwelling, large mammal fauna is a shadow of what it was in prehistory. Since the disappearance of the dinosaurs, almost every landmass has been home to a changing roll call of large mammals. Of all the large mammals, the herbivores have attained the greatest sizes, and this, along with thick skin, horns, tusks, and antlers, has given them a lot of protection from potential predators. However, evolution always finds a way, and over the last 50 million years or so, there have been at least four separate mammal groups that have evolved a weapon to dispatch large, thick-skinned prey. The weapon is the saber tooth, and we have already been introduced to two types of extinct cat that were able to kill their prey with massively modified canine teeth (see the entries "Saber Tooth Cat" and "Scimitar Cat" in chapter 5).

When South America was rafted away from the other landmasses that formed the super-continent of Gondwanaland, it carried an unusual assemblage of mammals quite distinct from the inhabitants of the other continents. There were the forerunners of the sloths, ant-eaters, and armadillos we know today as well as less familiar types. Along with Australia, South America was also a marsupial stronghold, and for a while, these pouched mammals were very successful predators on this southern continent. For much of the time, South America was isolated, and the only large predators were the marsupials and giant, flesh-eating birds. Evolution even shaped members of this marsupial stock into an animal very similar to the more familiar saber tooth cats. This animal was the pouch-knife, and it is a very enigmatic creature.

This animal was first described in 1934 by the paleontologist Elmer Riggs, of the Field Museum in Chicago, from two incomplete skeletons discovered in Argentina. In terms of size, the pouch-knife was probably as large as a jaguar, though it had shorter legs. The preserved skulls of this extinct marsupial have been slightly distorted by fossilization, but they, with fragments of unearthed skeletons, are still the only decent fossil evidence of the pouch-knife. It is amazing that the skull of the pouch-knife is so superficially similar to those of the saber tooth cats, even though marsupials and cats sit on very different branches of the mammalian family tree. Again, this is another excellent example of convergent evolution and goes to show how nature can come up with similar solutions to the same problem in very different locations.

The skulls of saber tooth cats and the pouch-knife may be very similar at first glance, but there are many major differences, which show that the pouch-knife was a very different mammal. Its sabers were enormous, relatively larger than those of *Smilodon populator*, and they also grew throughout the animal's life, which was very useful as the tips and cutting edge always remained sharp. As the pouch-knife's teeth grew continuously, they could not be fixed in the jaw with a bulbous anchor like those of the saber tooth cats. Instead, they grew from long roots that extended to a position well behind and above the pouch-knife's eyes. Also, when the mouth was closed, these massive canines were protected by scabbard-like outgrowths of the pouch-knife's chin. These scabbards were equipped with tough pads that may have sharpened the teeth as the jaws were opened and closed.

As the fossil record for the pouch-knife is so scant, we only have a very limited idea of how it lived. It seems that this pouched predator lived in a savannahlike environment, sharing this open habitat with the other strange denizens of South America, including the numerous types of large, native ungulate; the extinct relatives of the sloths and armadillos; numerous types of rodent (some of them huge); and the giant, predatory terror birds (see the entry later in this chapter). The pouch-knife was undoubtedly a predator as the canines are suited to killing and the shearlike cheek teeth are like those in the skull of a big cat—ideal for slicing flesh from a carcass. Not only was this extinct marsupial equipped with impressive teeth, but the region of the skull that once housed its hearing organs is well developed, indicating that this sense was probably acute. Along with sabers and a good sense of hearing, the pouch-knife's neck muscles and forelimbs must have been very strong. Powerful forelimbs allowed the marsupial to get a firm grip on prey, while the muscular neck allowed the stabbing canines to be driven through the tough hide of the victim into the soft tissues beneath. The hip joint of this animal is also very flexible, and some experts think it may have been capable of moving on its hind legs over short distances, much like the thylacine (see the entry in chapter 1). This may have been important in reaching up to the neck of its prey to deliver the killer bit. Exactly what prey the pouch-knife killed and ate is unknown, but it may have been a specialist predator of the numerous small- to medium-sized herbivores that once roamed South America. As it was short-legged and quite sturdy, it is doubtful that the pouch-knife was capable of pursuing its prey over any great distance. It probably opted for an ambush strategy, concealing itself behind pampas vegetation before it launched a lightning lunge at its victim. We may only be able to guess at the feeding behavior of this extinct predator, but we know much more about how it reproduced. As it was a marsupial, it probably had a pouch, and if the thylacine is a good example of a predatory marsupial, the female pouch-knife may have had a pouch that faced backward so that dirt and vegetation did not get into the furry pocket that cosseted her developing young. You can imagine a young pouch-knife, its sabers still small and developing, slipping from its mother's pouch to investigate the outside world.

The pouch-knife is a mysterious animal, and the fossil record of the group of animals to which it belongs is far from complete, but this is due to the fact that fossilization is very rare, and finding what's left of these long-dead animals is very difficult and often relies on sheer luck. What we do know is that the ancestors of the pouch-knife lived around 13 to 14 million years ago. What caused the demise of the pouch-knife? One unlikely theory is that an asteroid impact in South America caused the local extinction of many animal species, including the pouch-knife. There is some limited evidence for an impact event, but it is impossible to say if it was disastrous enough to kill off some of the South American fauna. It is more likely that the Great American Interchange led to the demise of the pouch-knife (see the "Extinction Insight" in chapter 2). This began around 3 million years ago as a result of the formation of the Isthmus of Panama—a land bridge that fully connected North and South America for the first time. Land and freshwater animals freely traversed this bridge, and the mammals of South America were exposed to an influx of North American animals. At the time of this event, the predatory marsupials were already on the decline, and we know from recent extinctions in Australia that when predatory marsupials come into direct competition with placental mammals, they often lose. The dwindling pouch-knife may have never been very abundant, and in their last few thousand years, these marsupials may have been pitted against the much larger saber tooth cats, which migrated into South America from the north. These felines may have been more efficient at dispatching their thick-skinned prey, contributing to the extinction of the pouch-knife.

- As the skull of the pouch-knife has been distorted by fossilization, the big canines are actually splayed, and it was once thought that this is how the living animal must have looked. This idea is now rejected as such large, splayed teeth jabbed into a victim would have generated skull-splitting force.
- In the marsupials we know today, the young become independent as soon as they finish taking their mother's milk. However, the pouch-knife young may have stayed with their mother for extended periods of time to learn and develop the specialized killing technique used by this species.
- Victorian paleontologists came up with all sorts of ideas for how the pouch-knife used its impressive teeth. One of the more amusing theories is that the marsupial used its canines and scabbards like can openers to open the domed carapaces of glyptodonts (see the entry in chapter 5). Even if a pouch-knife was foolish enough to gnaw the bony shell of one of these animals, it would have quickly found itself with a pair of broken canines.

Further Reading: Argot, C. "Evolution of South American Mammalian Predators (Borhyaenoidea): Anatomical and Palaeobiological Implications." *Zoological Journal of the Linnean Society* 140 (2004): 487–521.

TERROR BIRD

Terror Bird—This terror bird (*Titanis* sp.) skull is almost 50 cm long, and it clearly shows the massive bill, with its hooked tip, that was used to kill and dismember the unfortunate mammals of ancient South America. (Natural History Museum at Tring)

Terror Bird—This progressive reconstruction of a terror bird (*Paraphysornis brasiliensis*) demonstrates the powerful legs and robust skeleton of these remarkable birds. (Renata Cunha)

Scientific name: *Phorusrhacids*
Scientific classification:
> Phylum: Chordata
> Class: Aves
> Order: Gruiformes
> Family: Phorusrhacidae

When did it become extinct? Experts disagree on when the last terror bird became extinct. Some scientists argue that it was as little as 15,000 years ago, which is very unlikely. It's far more probable that they became extinct around 1.8 million years ago.

Where did it live? The remains of these animals have been found throughout South America, and the fossils of one species have been found in Florida and Texas.

In the right circumstances, birds can evolve into giants. In the vast majority of cases, they have done this on oceanic islands in the absence of any large land predators. Most of the extinct giant birds are decidedly lacking when it comes to predatory ferocity. Birds like the moa and elephant bird were big animals, but they were gentle vegetarians. However, a long-legged bird living in South America several million years ago gave rise to a group of birds collectively known as terror birds. As their name suggests, these animals were not the sort of feathered critter you would be pleased to see at your bird feeder. They were big birds; the smallest were at least 1 m tall, while the biggest stood as high as 3 m. All of them bear the hallmarks of being ferocious predators. Why these nightmarish birds came to evolve in South America is not fully understood as no other place on earth has ever produced a group of predatory giant birds. Gigantism in birds is normally associated with herbivory, yet whatever conditions prevailed in South America many millions of years ago allowed the evolution of a successful and varied group of feathered carnivores.

Following the extinction of the dinosaurs, many niches in earth's ecosystems were left wide open for the vertebrate survivors—the mammals, birds, and remaining reptiles—to evolve into, and for a while, apparently, the terror birds had a power struggle with the mammals for the dominance of the terrestrial ecosystems in South America. Many of them were big and powerful enough to have been the top predators at the time, and many mammals were definitely their prey.

All but one of the terror birds paleontologists know of today have been unearthed in South America. One species (*Titanis walleri*) managed to reach North America, and it appears to have been quite a success, surviving for more than 3 million years, until it disappeared around 1.8 million years ago—the last of its kind to become extinct. Even though this American species was not the biggest terror bird, it must have still been a terrifying animal. Its vital statistics are impressive: 1.4 to 2.5 m tall and 150 kg in weight. It also had an immense, hooked bill, and with such an impressive beak, it could have probably swallowed a lamb-sized animal in one gulp.

Although we can piece together the skeletons of the terror birds, it's impossible to know what their plumage was like. However, we can look at living birds for clues, and if the other flightless birds are anything to go by, the terror bird's feathers may have been rather hairlike. Like the vast majority of flightless birds, terror birds had stubby little wings, but what they lacked in the wing department they more than made up for with their long, powerful legs, which ended in large feet and fearsome claws. These legs gave these animals a good turn of speed, and it has been estimated that some species of terror bird could reach speeds of 100 km per hour—comparable to a cheetah. The combination of running, big talons, and a monstrous beak made the terror birds very effective predators. It is possible to imagine one of these birds snapping at the hooves of ancient mammals as it pursued them across the grasslands of the Americas. Smaller animals were probably immobilized with the sharp talons before being torn apart by the fearsome hooked bill or even swallowed whole after having their skull crushed in the bird's vicelike grip. Larger prey animals may have been disemboweled with kung fu–style kicks, and it is even possible that crushing kicks may have been used to crack the larger bones of big prey to get at the nutritious marrow within.

Even if the last terror bird became extinct around 1.8 million years ago, these were successful animals that, as a group, survived for more than 50 million years, some of them even taking on the mantle of top land predator in the ecosystems in which they lived. However, around 2.5 million years ago (during the Pliocene epoch), something happened that completely changed the course of life for South America's unique animals—the Great American Interchange (see the "Extinction Insight" in chapter 2). The land bridge that formed between North and South America, what is now known as the Isthmus of Panama, allowed animals from the north to migrate into South America. Among them were lots of predatory cats, and it has been proposed that these animals were so effective as predators that they outcompeted the terror birds. The talons and beaks of the terror birds were no match for the teeth, claws, and hunting prowess of the invaders from the north. This is a very neat answer for the cause of the extinction of the terror birds; however, the extinction of successful animals is very rarely due to one factor, but a combination of events. Perhaps climate change directly affected the terror birds by changing their habitats and the populations of their prey. Although there is a great deal we don't know about the life and times of the terror birds, we do know that one of their number somehow managed to cross into North America and spread through the southern states. For a long time, it was assumed that the North American terror bird spread north via the land bridge, but analysis of its ancient bones paints an alternative picture, as they appear to have reached the southern states of America before the land bridge formed. Perhaps falling sea levels, due to the growth of the polar ice sheets, revealed a path of island stepping-stones across the gap of open ocean that would become the Isthmus of Panama. These stepping-stones allowed the giant birds to colonize the prehistoric North America. Maybe other species of terror bird, the remains of which are as yet undiscovered, also reached North America before following the rest of their amazing kind into the pages of earth history.

+ The closest living relatives of the terror birds are the seriemas of South America. A bird similar to the living seriemas probably gave rise to the 17 species of terror bird that are known today from fossilized remains. These fossils cover a long period of geologic time, from about 60 million years ago to 1.8 million years ago, which goes to show how successful these birds were.
+ For many millions of years, large, carnivorous, placental mammals were absent from South America, and in the absence of these predators, the ancestors of the terror birds evolved to fill this niche.
+ The largest species of terror bird was the gargantuan *Brontornis burmeisteri*, identified from remains discovered in Argentina. This heavily built bird, with its massive head, rivals the elephant bird of Madagascar for the title of the biggest bird that has ever lived. Remains of this monster are very rare, but it has been estimated that it weighed 350 to 400 kg and was probably around 3 m tall. Like the rest of its kind, it was a meat eater, and in life, it must have been a truly spectacular creature.
+ In 2003, a high school student in Patagonia unearthed an almost complete skull of a new terror bird species and one that may have been even bigger than *B. burmeisteri*. This skull was not much less than 1 m long, and it gives a true sense of what imposing creatures the largest terror birds must have been.

Further Reading: Marshall, L.G. "The Terror Birds of South America." *Scientific American* 270 (1994): 90–95; Alvarenga, H.M.F., and E. Höfling. "A Systematic Revision of the Phorusrhacidae (Aves: Ralliformes)." *Papéis Avulsos De Zoologia* 43 (2003): 55–91; MacFadden, B.J., J. Labs-Hochstein, R.C. Hulbert, and J.A. Baskin. "Revised Age of the Late Neogene Terror Bird (*Titanis*) in North America during the Great American Interchange." *Geology* 35 (2007): 123–26.

GIANT HYENA

Scientific name: *Pachycrocuta brevirostris*
Scientific classification:
    Phylum: Chordata
    Class: Mammalia
    Order: Carnivora
    Family: Hyaenidae

Giant Hyena—The giant hyena was about the same size as a big lioness and was probably capable of dismembering some very large carcasses with its formidable teeth and jaws. (Renata Cunha)

When did it become extinct? The giant hyena is thought to have become extinct around 500,000 years ago.

Where did it live? The remains of this animal have been found in Africa, Europe, and all the way through Asia to China.

The spotted hyena is a beautifully adapted predator and scavenger of the African continent. These animals have a long evolutionary heritage of at least 70 species extending back at least 15 million years. The earliest known hyenas were mongoose-sized animals that were probably insectivorous or omnivorous, but over time, they evolved into specialized hunters and scavengers, the largest of which was the giant hyena.

In general appearance, the giant hyena was similar to the spotted hyena, only much bigger. It was a powerfully built animal, and a specimen in its prime probably weighed the same as a big lioness, around 150 kg, or possibly more (for comparison, a really big spotted hyena weighs around 90 kg). Due to its short legs, it was only marginally taller at the shoulder than a spotted hyena (about 1 m), and its big skull was equipped with some formidable teeth, very well suited to dismembering carcasses.

The spotted hyena is often portrayed as nothing but an idle, scavenging animal that depends on the kills made by lions and other cats for its food. It's true that the spotted hyena is certainly not above scavenging, but it is also a very accomplished predator, able to use teamwork to bring down antelopes and animals as large as zebras. What can we deduce about the life of the extinct giant hyena from the life of the spotted hyena? With its relatively short legs, the giant hyena was not built for long-distance pursuits like its living relative, but this animal was very much of its age, and some of the herbivores that fell prey to the carnivores of the Pleistocene were less fleet of foot than the ungulates of the African plains of today. We only have to look at the top predators that lived alongside the giant hyena: big scimitar cats and other large felines built for strength, not stamina. The giant hyena may have been able to catch its own prey, especially if it hunted in groups like the living spotted hyena, but scavenging in groups was probably its mainstay. A kill made by one of the many big cats of the day would have quickly attracted the attention of a group of giant hyenas. A cat like *Homotherium* probably defended its kill from one or two giant hyenas, but a bigger group of these scavengers was more of a problem. The bite of a giant hyena was very powerful, and a bad wound can be a death sentence for a predator; therefore the owner of the kill may have been forced to begrudgingly surrender the carcass to the hyena clan.

With the owner of a kill driven away, the giant hyena could do what it did best and fill its capacious stomach with meat, and use its bolt-cropper jaws to shear the bones of the carcass and carry certain choice cuts back to its lair, where cubs were probably waiting for food. In China, there is clear evidence of the giant hyena carrying food back to its lair. Zhoukoudian is a cave system near Beijing, and it is here that paleontologists found a great haul of mammal bones in the 1930s, including the remains of several giant hyenas. The hyenas had undoubtedly used these caves as lairs, and this is where they brought bits of carcasses to feed their growing cubs. Amazingly, the remains of at least 40 *Homo erectus* individuals were also unearthed in the caves. The question is, did our ancient ancestors live in these caves, or were their dismembered remains carried there by the giant hyenas? *Homo erectus* was definitely capable of making and using weapons (see the entry "*Homo erectus*" in chapter 6), but was this hominid capable of fending off a group of 150-kg bone breakers? Five hundred thousand years ago, our ancestors were on the menu for lots of different predators, and even if giant hyenas never hunted *Homo erectus* directly, the carcass of one of these hominids, killed by one of the big cats, was certainly big enough to arouse the interest of these scavengers.

The most recent known remains of the giant hyena are around 500,000 years old, but we have no firm date for when this species became extinct. We do know that the youngest fossils of the giant hyena correspond to a time when the earth was entering another of the glaciations that have punctuated the last 2 million years. The climate became drier and the verdant habitats available to the big herbivores dwindled. As their food disappeared, many of these megaherbivores disappeared, and so, too, did their predators, including some of the large cats. Primarily a scavenger, the giant hyena was dependent on these large predators for food, and as they disappeared, it, too, was doomed.

- The general appearance of hyenas suggests a close evolutionary link to the dog family; however, hyenas are an offshoot of the cat branch of the carnivores, and therefore they are more closely related to cats than dogs.

- In contrast to group-living felines, like lions, female spotted hyenas are the dominant sex, and each hyena clan is ruled by an alpha female. Taking charge has had some unusual effects on the female's anatomy as the increased levels of testosterone coursing through the blood of a female spotted hyena has led to the development of a false penis and scrotum. The pseudopenis is actually a hugely modified clitoris, which is erectile just like a real penis. The pseudoscrotum is formed from the exterior skin of the female genitals.
- Like our ancestor *Homo erectus*, the giant hyena evolved in Africa and then proceeded to disperse into Europe and Asia, reaching as far east as China.

Further Reading: Turner, A., and M. Antón. "The Giant Hyaena, *Pachycrocuta brevirostris* (Mammalia, Carnivora, Hyaenidae)." *GEOBIOS* 29 (1996): 455–68.

GIANT APE

Giant Ape—A giant ape shown alongside the silhouette of a modern human to give an idea of size. They may have been even larger than this, although it is not known if they were bipedal. (Phil Miller)

Scientific name: *Gigantopithecus blacki* and *G. giganteus*
Scientific classification:
   Phylum: Chordata
   Class: Mammalia
   Order: Primates
   Family: Hominidae
When did it become extinct? The giant apes are thought to have become extinct around 200,000 years ago.
Where did it live? The remains of *G. blacki* have been found in southern China and northern Vietnam, while the remains of *G. giganteus* have been found in northern India.

A visit to the Himalayas would not be complete without tales of yeti, the hairy, apelike creatures that are supposed to inhabit this immense mountain range. As long ago as the 1830s, explorers to these majestic mountains have returned with tales of this beast, tales that have captured the public's imagination. As there is no irrefutable proof of the yeti's existence, it will never be more than a yarn to scare mountaineers; however, 200,000 years ago, there were at least two species of giant ape that lived in Asia, though apart from their size, they bear little resemblance to the cryptozoological accounts that fire the imagination.

In 1935, the respected paleontologist Ralph von Koenigswald visited a traditional Chinese medicine shop and found the molars of what were undoubtedly a primate. Fossil teeth were coveted in Chinese medicine. Known as dragon's teeth, they were ground down into a powder for use in a variety of treatments. The teeth von Koenigswald found were saved from being crushed and were formally identified as coming from the mouth of an extinct primate. Since the discovery of these first teeth, other fossils of these primates have come to light, including more teeth and several jawbones from various cave sites. At the moment, this is all we have to go on, but paleontologists have put forward several ideas as to what these animals looked like and how they lived. In the same way that reconstructions of the giant shark have been produced from nothing more than teeth (see the earlier entry in this chapter), paleontologists have used the teeth and jawbones of these giant apes to build a picture of what the living creatures may have been like.

As their name suggests, the giant apes were large animals. Estimates for just how large they were vary, but some experts think that *G. blacki* (the larger of the two species) could have been 450 kg. As no leg bones of these animals have ever been found, we cannot say for sure exactly how they moved, though they most likely walked around on all fours like gorillas (*Gorilla gorilla*). If *G. blacki* were to rear up on its hind legs, it's estimated to have been over 3 m tall—a truly startling thing to imagine. Obviously, these estimates have to be treated with caution because all we have to go on are the teeth, and it is possible that they belonged to an ape with a disproportionately large head. If the size estimates of the giant apes are correct, they were the largest primates that have ever lived, and the largest species was more than twice the weight of the largest male gorilla. Like those of the gorillas, the molars of the giant apes appear to be suited to pulverizing plant food. It's believed that they made use of the forests of bamboo that grow in Southeast Asia, much in the same way as the living giant panda (*Ailuropoda melanoleuca*).

Most of the remains of the giant ape have been found in caves, but it is very unlikely that the living animal was a cave dweller. No primates, except humans, routinely frequent caves, so why have the remains of this extinct ape come to light in such situations? The answer is porcupines. These prickly animals will drag all manner of things back to their lair to gnaw on, and thousands of years ago, the bones of giant apes were among the things they collected. Porcupines and their love of gnawing is also the reason we find nothing more substantial than the teeth and jawbones of the giant apes. Porcupines gnawed at the limb bones and the other large pieces of the skeleton until there was nothing left, except the very hard enamel caps of the teeth and the compact bone of the mandible.

The most recent remains of the giant apes are around 200,000 years old, and there is currently no evidence as to how or exactly when they died out. Regardless of exactly when these giant primates died out, our ancient ancestors *Homo erectus*, who had reached as far east as Indonesia at least 840,000 years ago, may have come into contact with them. Their reaction to these animals is hard to imagine, but if the giant apes were gentle plant eaters, they could have been just another animal to kill and eat.

One thing is certain: the bones of this animal are very rare, but it's hopefully only a matter of time before more complete remains are unearthed to give us a better idea of how this animal looked and when it vanished.

+ The yeti is known by many names, including the "abominable snowman," a name that was undoubtedly coined by British explorers in the nineteenth century. There are monasteries in Nepal that treasure the supposed remains of the yeti, including a scalp and the bones of a hand. Tests have been conducted on the scalp, and the skin is actually from a goat.
+ The tales of the yeti are not the only stories of giant primates. There are reports of large bipedal primates from other parts of the world, the most familiar of which is the Sasquatch (bigfoot) of North America. The world is certainly huge, with many remote places, but is it big enough to hide viable populations of 300- to 500-kg primates during more than 500 years of intense exploration? As the bones of the giant ape testify, the earth, at some point, has been home to huge primates, but the chances of them surviving into the modern day, amid more than 6 billion humans, are vanishingly small. Stories of the Sasquatch and yeti undoubtedly capture the public's interest, but the stark realization is that they are probably nothing more than figments of the imagination.

Further Reading: Simons, E. L., and P. C. Ettel. "*Gigantopithecus.*" *Scientific American*, January 1970; Ciochon, R. L., J. Olsen, and J. James. *Other Origins: The Search for the Giant Ape in Human Prehistory*. New York: Bantam Books, 1990.

GIANT CAMEL

Scientific name: *Titanotylopus nebraskensis*
Scientific classification:
   Phylum: Chordata
   Class: Mammalia
   Order: Artiodactyla
   Family: Camelidae

Giant Camel—Compared to the modern camel, the giant camel was enormous. It stood around 3.5 m at the shoulder. (Phil Miller)

When did it become extinct? The giant camel is thought to have become extinct around 1 million years ago.

Where did it live? The giant camel lived in North America.

It is difficult to use the word *camel* without picturing the deserts of the Middle East and Asia; however, it may come as a surprise to learn that the camels originated and underwent most of their evolution in North America. The oldest ancestors of the camels are rabbit-sized, four-toed animals known from 40-million-year-old fossils. Over millennia, these ancestors gave rise to a number of species, of which only a few survive today. The giant camel was one of these species, and a very large one at that. The most familiar camel alive today, the dromedary (*Camelus dromedarius*), can be 2.1 m at the shoulder, 3 m long, and weigh 1,000 kg—a big animal, but it would look puny next to the giant camel, which, at 3.5 m tall and at least 1,800 kg, was the biggest camel that has ever lived.

Camels are very interesting animals that have evolved a number of adaptations for surviving in very tough environments, and the two species of camel alive today, the dromedary and the Bactrian (*Camelus bactrianus*), are born survivors, able to thrive in some of the harshest places on earth. What do we know about the giant camel? Was it similarly hardy? In some ways, it may have been, but the America in which it lived was very different to the land we know today. The climate was warmer and moister, so it is unlikely the giant camel was as hardy as the living species.

Camels are unique for their humps, which at one time were thought to store water, but are now known to store fat, making it possible for these animals to go for long periods of time without food. There is no way of knowing if the giant camel was humped. The vertebrae of its back do bear long spines, just like those of the modern camels, but this may have been for the attachment of the nuchal ligament that holds the head up. Camels also have a battery of adaptations that enable them to survive without water for several days at a time. They can lose up to 25 percent of their body weight in moisture before they get into difficulties. In contrast, most other mammals die if they lose only 3 to 4 percent. The camel limits the moisture it loses in its breath and produces viscous urine, both of which cut down on water loss. Dehydration in other mammals results in the blood getting progressively thicker, straining the heart until it can no longer beat effectively, but the camels get around this problem with red blood cells that are oval, rather than round, and it is thought that this enables the camel's blood to keep flowing even when the animal is dehydrated. When camels do find water, they really make up for their hardships, and they quench their thirst by drinking around 100 liters in one go, some of which is stored in special cavities in the lining of their large stomach.

It is unlikely that the giant camel was similarly equipped for survival. The America in which it lived, 1 to 5 million years ago, was a very different place to the continent we know today, and much of the land was forested, albeit sparsely in places. The giant camel probably never had to go without water for days at a time, but it needed the means of making the best use of the vegetation the open American forests provided. Its digestive system was undoubtedly very similar to that of the living camels, employing symbiotic bacteria to digest tough plant food. The giant camel, like its living relatives, could probably tolerate massive ranges in temperature that would cause most other mammals to keel over. Their thick fur can insulate them from the cold and the heat of the sun, enabling them to survive in temperatures as low as −40 degrees Celsius and as high as +40 degrees Celsius. The forested plains of Nebraska 1 to 5 million years ago were much warmer than today, but winter temperatures can be still be very low in the middle of a large continent, so the giant camel must have coped with cold winter conditions.

Although camels are champion survivors, they can be quite short-tempered beasts, and it seems the giant camel was no exception. Males of this extinct species sported well-developed canine teeth, and it is very likely that they used these to good effect during the breeding season, when disputes with other males over territory and females were commonplace.

The youngest remains of the giant camel are about 1 million years old, and we know that there were no humans in North America to hunt them at that time, so why did they become extinct? We don't know for sure, but climate change was the likely culprit. As the climate cooled, the preferred habitat of the giant camel—open forest—may have been replaced by grassland, and this enormous beast was squeezed out of existence.

+ Today, the camels and their relatives are represented by the dromedary and Bactrian camels of the Old World and the llama (*Lama glama*), guanaco (*Lama guanicoe*), vicuña (*Vicugna vicugna*), and alpaca (*Vicugna pacos*) of the New World.
+ Even though the camels, as a group, originated and underwent most of their evolution in North America, they died out there about 10,000 years ago, but millions of years ago, the ancestors of the two living camel species migrated into Asia via the Bering land bridge.
+ The dromedary camel is actually extinct in the wild. It was domesticated at least 3,500 years ago (possibly as much as 6,000 years ago) and proved so useful to early

civilizations that its populations exploded, and the wild animals were tamed or bred out of existence. The Bactrian camel still exists in the wild, but the population is no more than 1,000 animals, and they are limited to the northwestern corner of China and Mongolia, where they manage to survive in the unbelievably hostile Gobi Desert.

+ The camels use a pacing gait to get around. The legs on the left side of the body step together, followed by the right legs. This unusual gait may look awkward, but it is actually a very energy-efficient way of getting around. A camel's pace can be quite unstable because of all the side-to-side motion; however, this is counteracted by its well-developed footpads.

Further Reading: Harrison, J. A. "Giant Camels from the Cenozoic of North America." *Smithsonian Contributions to Paleobiology* 57 (1985): 1–29; Breyer, J. "*Titanotylopus* (= *Gigantocamelus*) from the Great Plains Cenozoic." *Journal of Paleontology* 50 (1976): 783–88.

GLOSSARY

Amphibian—an animal that spends its time in the water and on land.

Amplexus—the name given to the reproductive embrace of frogs and toads.

Apatite—the mineral that forms the enamel of teeth and reinforces bone.

Archipelago—a group of islands.

Articulate—the way in which bones are arranged in an animal's skeleton.

Artiodactyls—the group of herbivorous animals that includes deer, sheep, and cattle characterized by their cloven hooves.

Asphalt—the dark, sticky substance that remains of crude oil after the light, volatile fractions have evaporated.

Asteroid—a lump of orbiting rock left over from the formation of the star systems, some of which can be several kilometers across.

Basalt—a type of fine-grained igneous rock.

Bering land bridge—a large tract of land that connected Asia to North America. Rising sea levels at the end of the last glaciation flooded this land bridge.

Cambrian—one of the earth's geological ages, which extended from 490 to 543 million years ago.

Canid—a name for the group of predatory mammals commonly known as dogs.

Carboniferous—one of the earth's geological ages, which extended from 299 to 352 million years ago.

Carrion—the name given to dead and decaying animals that are eaten by scavengers.

Cellulose—the glucose-based polysaccharide that is found in the cell wall of all plants.

Centra—a disc-shaped section of the vertebral column.

Chytrid fungi—a type of fungi that infects the soft skin of amphibians, leaving them open to other opportunistic infections.

Cloaca—the common opening for the genital, urinary, and digestive tract that is found in all fish, amphibians, reptiles, birds, and monotreme mammals.

Cloning—the technique of producing an exact copy of an animal from the DNA inside one of its cells.

Cloud forest—forest growing in mountainous areas that is often shrouded in cloud.

Coccoliths—individual calcium carbonate plates from the shell that surround certain kinds of single-celled algae.

Comb—the fleshy protuberances on the head of certain birds.

Continental drift—the process by which the continental plates move around on the lava that forms the earth's mantle.

Convergent evolution—in evolutionary biology, the process whereby organisms not closely related independently evolve similar traits as a result of having to adapt to similar environments or ecological niches.

Cretaceous—one of the earth's geological ages, which extended from 65 to 145 million years ago.

Cro-Magnon—a term usually used to describe the oldest modern humans of Europe.

Cuticle—the nonmineral outer covering of an organism.

Devonian—one of the earth's geological ages, which extended from 359 to 416 million years ago.

Dinosaur—a group of reptiles that dominated terrestrial ecosystems for about 160 million years until the end of the Cretaceous.

Ecosystem—a system formed by the interaction of a community of organisms with their environment.

Endemic—an organism exclusively native to a certain place.

Endocast—the replica of a brain that is formed when sediments or other materials fill the buried cranium of a dead animal.

Eurasia—the landmass comprising Europe and Asia.

Fauna—the animal life in an ecosystem.

Femur—in all vertebrates with legs, the bone between the hip and knee.

Firn—ice that is at an intermediate stage between snow and glacial ice.

Flora—the plant life in an ecosystem.

Folk memory—stories that are passed, orally, from one generation to the next.

Foraminifera—tiny, single-celled organisms, often shelled, that live in profusion in the oceans.

Gizzard—the muscular organ found in the digestive tract of birds and other animals that grinds up food.

Gondwanaland—a probable landmass in the Southern Hemisphere that separated many millions of years ago to form South America, Africa, Antarctica, and Australia.

Greenhouse gas—any gas in the atmosphere that traps the heat reflected from the earth's surface, e.g., carbon dioxide or methane.

Holocene—the present geological epoch, which began around 10,000 years ago.

Hominid—a collective term for extinct and extant humans, chimpanzees, gorillas, and orangutans.

Interstadial—a period of colder temperatures during an interglacial.

Invertebrate—any animal that lacks a vertebral column.

Iridium—a very dense metallic element that is rare on earth but more common in asteroids and meteorites.

Island rule—a principle in evolutionary biology stating that members of a species get smaller or bigger depending on the resources available in the environment.

Joey—the name given to an infant kangaroo or wallaby.

Jurassic—one of the earth's geological ages, which extended from 145 to 199 million years ago.

Keel—the large extension of the sternum (breastbone) that serves as a muscle attachment in all flying birds.

Kelp—large seaweeds often found growing in so-called forests in shallow, nutrient-rich waters.

Keratin—a structural protein that is found in skin, hair, hooves, and claws.

Laurasia—the northern part of the Pangaean supercontinent comprising Asia, Europe, and North America.

Magma—molten rock that sometimes forms beneath the surface of the earth.

Malagasy—anything related to the island of Madagascar, including the people and the language.

Mantle—the 2,900 km thick layer of the earth's interior that surrounds the core.

Māori—the indigenous Polynesian people of New Zealand and their language.

Marsupial—a group of mammals native to Australasia and South America that give birth when the young are in a very early stage of development; the remainder of their development takes place in a pouch.

Megafauna—any species of large animals, but often used to refer to the large mammals that have become extinct in relatively recent times.

Micropaleontologist—a paleontologist who studies microfossils.

Mineralization—the process by which an organic substance is converted into an inorganic one.

Miocene—a geological epoch that extended from 5.3 to 23 million years ago.

New World—the Western Hemisphere, which includes the Americas.

Niche—the way in which an organism makes a living in a habitat.

Nymph—the immature stage of an insect that does not go through metamorphosis.

Old World—Europe, Africa, and Asia.

Ordovician—one of the earth's geological ages, which extended from 443 to 488 million years ago.

Osteoarthritis—a degenerative disease of the joints.

Osteomyelitis—a bacterial infection of the bone or bone marrow.

Ozone layer—the layer of ozone gas high in the atmosphere that absorbs some of the potentially damaging ultraviolet radiation in sunlight.

Paleoanthropologist—a scientist who studies ancient humans.

Paleontologist—a scientist who studies prehistoric life forms on earth through the examination of fossils.

Pangaea—the supercontinent comprising all the earth's landmasses that existed about 250 million years ago.

Pelage—the coat of a mammal, consisting of hair, fur, wool, or other soft covering, as distinct from bare skin.

Permafrost—soil that is at or below the freezing point of water for two or more years.

Permian—one of the earth's geological ages, which extended from 251 to 299 million years ago.

Perrisodactyls—the group of herbivorous animals that includes horses, rhinoceri, and so on, characterized by their odd number of hooves.

Photosynthesis—the process by which plants use sunlight to convert carbon dioxide and water into food.

Pinniped—the group of mammals that includes seals and sea lions.

Plankton—the mass of passively floating, drifting, or somewhat motile organisms occurring in a body of water, primarily comprising microscopic algae, protozoa, and the larvae of larger animals.

Pleistocene—a geological epoch that extended from 10,000 to 1.8 million years ago.

Pliocene—a geological epoch that extended from 1.8 to 5.3 million years ago.

Polynesian—the people and the culture originating from a group of around 1,000 islands in the Pacific Ocean.

Puggle—the name given to the young of echidnas.

Radiocarbon dating—a method for measuring the age of items containing carbon that is based on the steady decay of the radioactive carbon isotope, carbon-14.

Rainforest—forests characterized by high rainfall, typically 1,750 to 2,000 mm per year.

Sebaceous gland—small glands in the hair follicles of mammalian skin that secrete an oily substance known as sebum, which lubricates and protects the skin and hair.

Selective breeding—part of the domestication process by which animals and plants with useful traits are used for breeding to produce distinct breeds or cultivars.

Sexual selection—a theory that states that certain traits and characteristics can be explained by competition between members of a species.

Silurian—one of the earth's geological ages, which extended from 416 to 443 million years ago.

Sinkhole—a depression or hole in the surface of the ground due to the removal of soil and bedrock, usually by water.

Steppe—a grassland plain without trees.

Symbiotic—the close and often long-term relationship that exists between two species of organism.

Synapsid—the class of animals that includes mammals and their closest relatives such as the now extinct mammallike reptiles.

Syphilis—a sexually transmitted disease caused by a spirochete bacteria that afflicts many types of mammal.

Tertiary—the geological period that extended from 1.8 to 65 million years ago.

Thermal inertia—the ability of a substance to store internal energy as heat and has important implications for animals. A large animal has a lower surface area to volume ratio than a smaller animal; therefore it heats up and cools down at a slower rate.

Traps—the steplike landscape that can be found in regions of eroded flood basalt.

Triassic—one of the earth's geological ages, which extended from 199 to 251 million years ago.

Tuberculosis—a bacterial infection that affects many tissues in the mammalian body, sometimes leaving scars on the body and bones.

Tundra—treeless plains found in the extreme north and south and in mountainous areas, where plant growth is impeded by low temperatures and a short growing season.

Ultraviolet radiation—part of the spectrum of sunlight that is damaging to living things but that is absorbed by the ozone layer.

Vertebrae—the individual bones in the vertebral column of a vertebrate.

Vertebrate—any animal with a vertebral column.

Zoologist—a scientist who studies animals.

SELECTED BIBLIOGRAPHY

Agenbroad, L. D., and J. I. Mead. *The Hot Springs Mammoth Site: A Decade of Field and Laboratory Research in Paleontology, Geology, and Paleoecology.* Hot Springs, SD: The Mammoth Site of Hot Springs, South Dakota, 1994.

Agustí, J., and M. Antón. *Mammoths, Sabertooths, and Hominids: 65 Million Years of Mammalian Evolution in Europe.* New York: Columbia University Press, 2005.

Anderson, A. *Prodigious Birds: Moas and Moa-Hunting in New Zealand.* Cambridge: Cambridge University Press, 2003.

Archer, M., S. J. Hand, and H. Godthelp. *Australia's Lost World: Prehistoric Animals of Riversleigh.* Bloomington: Indiana University Press, 2001.

Barton, M., I. Gray, A. White, N. Bean, and S. Dunleavy. *Prehistoric America: A Journey through the Ice Age and Beyond.* New Haven, CT: Yale University Press, 2003.

Benton, M. J. *Vertebrate Palaeontology.* London: Wiley-Blackwell, 2004.

Cracraft, J., and F. T. Grifo. *The Living Planet in Crisis.* New York: Columbia University Press, 1999.

Day, D. *The Doomsday Book of Animals.* London: Ebury Press, 1981.

Diamond, J. A. *Collapse: How Societies Choose to Fail or Succeed.* New York: Viking Books, 2005.

———. *Guns, Germs, and Steel: The Fates of Human Societies.* New York: W. W. Norton, 1997.

———. *The Third Chimpanzee: The Evolution and Future of the Human Animal.* New York: HarperCollins, 2006.

Flannery, T. *The Future Eaters: An Ecological History of the Australasian Lands and People.* New York: Grove Press, 2002.

Flannery, T., and P. Schouten. *A Gap in Nature: Discovering the World's Extinct Animals.* New York: Atlantic Monthly Press, 2001.

Fuller, E. *Extinct Birds.* Sacramento, CA: Comstock, 2001.

Garbutt, N. *Mammals of Madagascar: A Complete Guide.* London: A&C Black, 2007.

Haines, T., and P. Chambers. *The Complete Guide to Prehistoric Life.* London: BBC Books, 2005.

Hallam, T. *Catastrophes and Lesser Calamities: The Causes of Mass Extinctions.* Oxford: Oxford University Press, 2004.

Lange, I., and D. S. Norton. *Ice Age Mammals of North America.* Missoula, MT: Mountain Press, 2002.

Lister, A., and P. Bahn. *Mammoths: Giants of the Ice Age.* Berkeley: University of California Press, 2007.

Long, J., M. Archer, T. Flannery, and S. Hand. *Prehistoric Mammals of Australia and New Guinea: One Hundred Million Years of Evolution.* Sydney: UNSW Press, 2002.

Macdonald, D. *The New Encyclopaedia of Mammals.* Oxford: Oxford University Press, 2001.

———. *The Velvet Claw: Natural History of the Carnivores.* London: BBC Books, 1992.

MacPhee, R.D.E., ed. *Extinctions in Near Time.* New York: Kluwer Academic / Plenum, 1999.

Martin, P.S. *Quaternary Extinctions: A Prehistoric Revolution.* Tucson: University of Arizona Press, 1989.

———. *Twilight of the Mammoths: Ice Age Extinctions and the Rewilding of America.* Berkeley: University of California Press, 2007.

Morwood, M., and P. van Oosterzee. *A New Human: The Startling Discovery and Strange Story of the "Hobbits" of Flores, Indonesia.* London: HarperCollins, 2007.

Murray, P., and P.V. Rich. *Magnificent Mihirungs: The Colossal Flightless Birds of the Australian Dreamtime.* Bloomington: Indiana University Press, 2003.

Nowak, R.M., ed. *Walkers Mammals of the World.* 6th ed. Baltimore: Johns Hopkins University Press, 1999.

Owen, D. *Tasmanian Tiger: The Tragic Tale of How the World Lost Its Most Mysterious Predator.* Baltimore: Johns Hopkins University Press, 2005.

Prothero, D.R. *After the Dinosaurs: The Age of Mammals.* Bloomington: Indiana University Press, 2006.

Quammen, D. *The Song of the Dodo: Island Biogeography in an Age of Extinction.* New York: Scribner, 1997.

Quirk, S., M. Archer, and P. Schouten. *Prehistoric Animals of Australia.* Sydney: Australian Museum, 1983.

Rich, P.V., and T.H. Rich. *Wildlife of Gondwana: Dinosaurs and Other Vertebrates from the Ancient Supercontinent.* Bloomington: Indiana University Press, 1999.

Rich, P.V., and G.F. van Tets. *Kadimakara: Extinct Vertebrates of Australia.* Princeton, NJ: Princeton University Press, 1991.

Strahan, R. *The Mammals of Australia.* Sydney, Australia: Reed Books, 1996.

Stringer, C., and P. Andrews. *The Complete World of Human Evolution.* London: Thames and Hudson, 2005.

Tricas, T.C., K. Deacon, P. Last, J.E. McCosker, T.I. Walker, and T. Leighton. *Sharks and Rays.* London: HarperCollins, 1997.

Turner, A., and M. Antón. *The Big Cats and Their Fossil Relatives.* New York: Columbia University Press, 1997.

———. *National Geographic Book of Prehistoric Mammals.* Washington, DC: National Geographic Society, 2004.

Woods, C.A., ed. *Biogeography of the West Indies: Past, Present, and Future.* Gainesville, FL: Sandhill Crane Press, 1989.

Worthy, T.H., and R.N. Holdaway. *The Lost World of the Moa: Prehistoric Life of New Zealand.* Bloomington: Indiana University Press, 2002.

Zimmer, C. *Smithsonian Intimate Guide to Human Origins.* London: HarperCollins, 2005.

SELECTED MUSEUMS IN THE UNITED STATES, CANADA, AND WORLDWIDE

American Museum of Natural History
Central Park West at 79th Street
New York, NY 10024-5192
USA
http://www.amnh.org

Museum of Paleontology
University of California
1101 Valley Life Sciences Building
Berkeley, CA 94720-4780
USA
http://www.ucmp.berkeley.edu

The Academy of Natural Sciences
1900 Benjamin Franklin Parkway
Philadelphia, PA 19103
USA
http://www.ansp.org

Carnegie Museum of Natural History
4400 Forbes Avenue
Pittsburgh, PA 15213
USA
http://www.carnegiemnh.org

The Field Museum
1400 South Lake Shore Drive
Chicago, IL 60605-2496
USA
http://www.fieldmuseum.org

Smithsonian National Museum of Natural History
10th Street and Constitution Avenue, NW
Washington, DC 20560
USA
http://www.mnh.si.edu

Page Museum at the La Brea Tar Pits
5801 Wilshire Boulevard
Los Angeles, CA 90036
USA
http://www.tarpits.org

Museum of Paleontology
The University of Michigan
1109 Geddes Avenue
Ann Arbor, MI 48109-1079
USA
http://www.lsa.umich.edu/exhibitmuseum

Peabody Museum of Natural History
Yale University
170 Whitney Avenue
New Haven, CT 06520-8118
USA
http://www.peabody.yale.edu

Natural History Museum of Los Angeles County
900 Exposition Boulevard
Los Angeles, CA 90007
USA
http://www.nhm.org

The Mammoth Site
1800 Highway 18 Truck Route
Hot Springs, SD 57747
USA
http://www.mammothsite.com/

The University of Alaska Museum of the North
907 Yukon Drive
Fairbanks, AK 99775
USA
http://www.uaf.edu/museum

Yukon Beringia Interpretive Centre
Whitehorse, Yukon Y1A 2C6
Canada
http://www.beringia.com

Natural History Museum
Cromwell Road
London SW7 5BD
UK
http://www.nhm.ac.uk

Natural History Museum at Tring
The Walter Rothschild Building
Akeman Street
Tring, Hertfordshire HP23 6AP
UK
http://www.nhm.ac.uk/tring

University Museum of Zoology
New Museums Site
Downing Street
Cambridge, Cambs CB2 3EJ
UK
http://www.zoo.cam.ac.uk/museum

The Moravia Museum
Zelny trh 6
659 37 Brno
Czech Republic
http://www.mzm.cz

Museum National d'Histoire Naturelle
Galeries de Paléontologie et d'Anatomie comparée
Paris 5ème
Jardin des Plantes 2, rue Buffon
Paris
France
http://www.mnhn.fr

Naturhistorisches Museum
Bernastrasse 15
CH-3005 Bern
Switzerland
http://www.nmbe.unibe.ch

Australian Museum
6 College Street (opposite Hyde Park)
Sydney, NSW 2010
Australia
http://www.austmus.gov.au

Museum Victoria
11 Nicholson Street
Carlton, Melbourne
Australia
http://www.museumvictoria.com.au

Museum of Natural History
ul. Św. Sebastiana 9
31-049 Kraków
Poland
http://www.isez.pan.krakow.pl

Finnish Museum of Natural History
Zoological Museum
Pohjoinen Rautatiekatu 13
Helsinki
Finland
http://www.fmnh.helsinki.fi

INDEX

About the Author

ROSS PIPER is an independent scholar. His lifelong interest in natural history, especially animals, led to academia and he went on to gain a first-class degree in zoology from the University of Wales, Bangor, and a PhD in entomology from the University of Leeds. He currently lives in Hertfordshire, England. This is his sixth book.